Routledge Revivals

Sir Isaac Newton

Originally published in 1927 this book presents the main features of Newton's life and his chief contributions to scientific knowledge. It gives the non-scientist, as well as the specialist, an insight into the life, personality and achievements of one of England's greatest scientists and polymaths.

Sir Isaac Newton
A Brief Account of his Life and Work

S.Brodetsky

First published in 1927 and as a second edition in 1929 by Methuen & Co. Ltd
This edition first published in 2024 by Routledge
4 Park Square, Milton Park, Abingdon, Oxon, OX14 4RN
and by Routledge
605 Third Avenue, New York, NY 10158.

Routledge is an imprint of the Taylor & Francis Group, an informa business

© 1929 S.Brodetsky.

The right of S.Brodetsky to be identified as the author of this work has been asserted by him in accordance with sections 77 and 78 of the Copyright, Designs and Patents Act 1988.

All rights reserved. No part of this book may be reprinted or reproduced or utilised in any form or by any electronic, mechanical, or other means, now known or hereafter invented, including photocopying and recording, or in any information storage or retrieval system, without permission in writing from the publishers.

ISBN 13: 978-1-032-94129-5 (hbk)
ISBN 13: 978-1-003-56914-5 (ebk)
ISBN 13: 978-1-032-94145-5 (pbk)
Book DOI 10.4324/9781003569145

SIR ISAAC NEWTON

From a copy of a bust by Roubilliac, in the possession of Christopher Turnor, Esq., at Stoke Rochford

SIR ISAAC NEWTON
A BRIEF ACCOUNT OF HIS LIFE AND WORK

BY

S. BRODETSKY
M.A., F.R.A.S.

PROFESSOR OF APPLIED MATHEMATICS, UNIVERSITY OF LEEDS

WITH A PORTRAIT, A
MAP AND 10 DIAGRAMS

SECOND EDITION

METHUEN & CO. LTD.
36 ESSEX STREET W.C.
LONDON

First Published . . . March 31st, 1927
Second Edition . . . 1929

Printed in Great Britain.

INSCRIBED TO THE MEMORY OF
MY MOTHER

PREFACE

THIS brief account of the life and work of Sir Isaac Newton does not claim to be a critical biography. The author's object has been to present the main features of Newton's life and his chief contributions to knowledge, in a manner that will be understood by a reader who possesses a very moderate grounding in the elements of science. Far too little attention is paid nowadays to the heroes of science, and it is hoped that the present volume will help to give to considerable numbers of schoolboys and schoolgirls, as well as to children of a larger growth, some insight into the life, personality and achievements of the greatest man of science England has produced.

Needless to say the author is indebted to many standard books, especially those by Sir David Brewster, and to a large number of scattered articles on Newton and his work. His thanks are due to Mr. Christopher Turnor of Stoke Rochford for kind permission to use the marble copy of the famous bust by Roubilliac for the frontispiece.

S. B.

LEEDS
February, 1927

PREFACE TO THE SECOND EDITION

THE very kind reception accorded to this book encourages the author in the belief that it has filled a definite gap in the literature concerning Sir Isaac Newton. No change has therefore been made in the general character of the book, but a number of corrections and slight alterations have been introduced. The author desires to express his sincerest thanks to a number of readers, who have very kindly drawn his attention to some errors of fact and phraseology.

S.B.

LEEDS
November, 1928

CONTENTS

CHAPTER		PAGE
I.	THE LINCOLNSHIRE CHILD, 1642–1661	1
II.	THE CAMBRIDGE STUDENT, 1661–1665	13
III.	THE DAWN OF MATHEMATICAL DISCOVERY, 1665	21
IV.	THE GERM OF UNIVERSAL GRAVITATION, 1666	29
V.	THE ANALYSIS OF LIGHT AND COLOUR, 1666	53
VI.	THE OPTICAL DECADE, 1668–1678	62
VII.	THE GRAVITATIONAL DECADE, 1678–1687	81
VIII.	THE "PRINCIPIA," 1687	101
IX.	THE TRANSITIONAL DECADE, 1687–1696	122
X.	THE GUARDIAN OF THE NATION'S COINAGE, 1696–1727	130
XI.	THE DOYEN OF BRITISH SCIENCE, 1703–1727	137
XII.	THE END, 1727	153

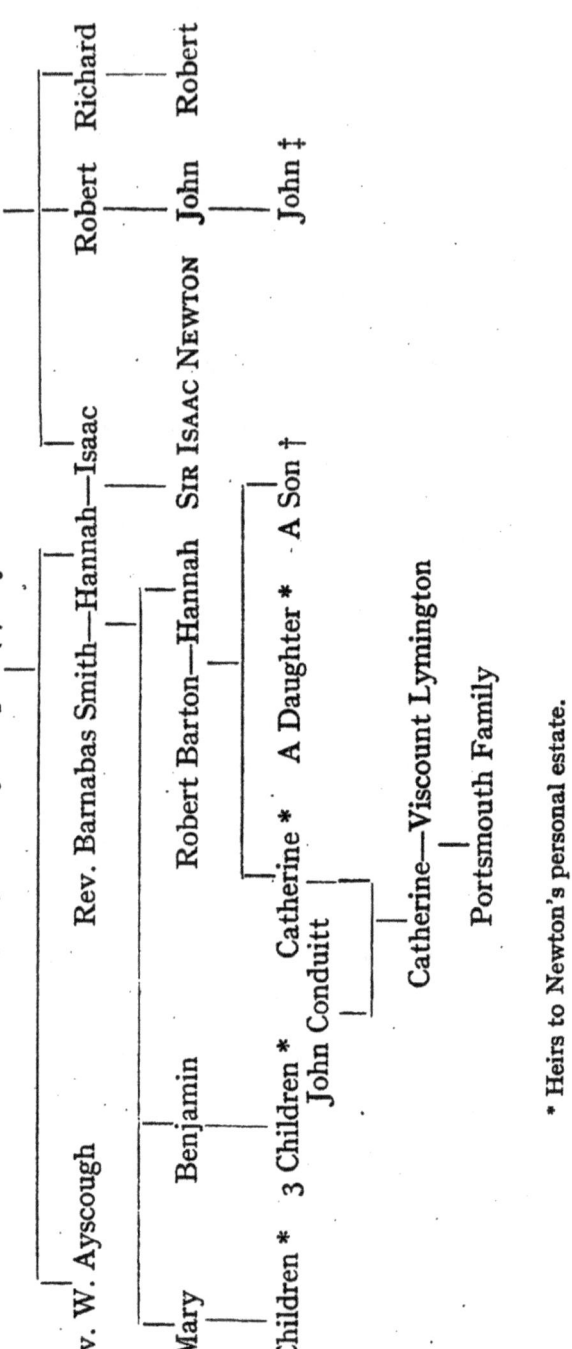

LIST OF ILLUSTRATIONS

SIR ISAAC NEWTON . . . *Frontispiece*
From a copy of a bust by Roubilliac, in the possession of Christopher Turnor, Esq., at Stoke Rochford

FIGURE		PAGE
	MAP OF NEWTON'S PART OF LINCOLNSHIRE .	6
	From a drawing by A. E. Taylor	
1.	AREA OF CIRCLE	24
2.	KEPLER'S LAWS	40
3.	CENTRAL ACCELERATION IN CIRCLE . .	49
4.	SPHERICAL ABERRATION	55
5.	SPECTRUM	59
6.	NEWTON'S RINGS	75
7.	FALLING BODY AND ROTATING EARTH . .	85
8.	CAUSE OF PRECESSION	115
9.	TIDES	116
10.	PATH OF PERIODIC COMET . . .	117

When Newton saw an apple fall, he found
   In that slight startle from his contemplation—
'Tis said (for I'll not answer above ground
   For any sage's creed or calculation)—
A mode of proving that the earth turn'd round
   In a most natural whirl, called " gravitation " ;
And this is the sole mortal who could grapple,
Since Adam, with a fall, or with an apple.

Man fell with apples, and with apples rose,
   If this be true ; for we must deem the mode
In which Sir Isaac Newton could disclose
   Through the then unpaved stars the turnpike road,
A thing to counterbalance human woes :
   For ever since immortal man hath glow'd
With all kinds of mechanics, and full soon
Steam-engines will conduct him to the moon.

                        LORD BYRON : " Don Juan,"
                            Canto the Tenth, I and II.

SIR ISAAC NEWTON

CHAPTER I

THE LINCOLNSHIRE CHILD, 1642–1661

> At first, the Infant,
> Mewling and puking in the nurse's arms.
> And then, the whining School-boy, with his satchel,
> And shining morning face, creeping like snail
> Unwillingly to school. And then, the Lover,
> Sighing like furnace, with a woful ballad
> Made to his mistress' eye-brow.
> WILLIAM SHAKESPEARE: "As You Like It."

IN August, 1642, the sword was drawn in the armed struggle between King Charles I and his Parliament. For nearly twenty years the struggle raged : first the King suffered complete defeat and Parliament became the supreme arbiter of the destinies of the nation ; then Parliament's power waned and became completely subject to the royal will of Charles II, restored to the powers and prerogatives of his less fortunate father. While the British people was in this oscillatory fashion hammering out a form of political and social life to express concretely its peculiar genius for ordered and law-abiding freedom, fully and jealously maintained personal liberty and inviolability combined with a dignified acceptance of, and submission to, laws derived by common consent, there was growing up in its midst a youth of unparalleled genius, who was to serve humanity in the scientific sphere, but whose life marked an epoch in human development as important and as significant as any marked by the career of dictator or liberator.

Detached from the great stream of agitated life, in the tiny hamlet of Woolsthorpe, in the parish of Colsterworth in Lincolnshire, about six miles south of Grantham, and a short distance from the main north road between London and York, there was born in the year of strife 1642 a frail child to a young and widowed mother. This child was Isaac Newton. When Charles I was executed by his indignant subjects and the monarchy gave way to a military commonwealth, this little boy was learning in primitive country schools the first elements of what was considered to be a suitable education for a small farmer. When Cromwell had expelled the Long Parliament and had been declared Lord High Protector young Newton went to Grantham in order to continue his education at the King's School. Cromwell died about the time when Newton's family had become convinced that he would never succeed as a farmer and as manager of the paternal estate in Woolsthorpe. In 1660, the year of the Stuart restoration and the complete subjection of the nation's will to the prerogative of its refound royal idol, Newton was preparing for entry upon his studies at Cambridge. The mind that was to set free the human intellect for unparalleled progress in the elucidation of the secrets of nature was approaching maturity, and was about to embark on a great adventure of discovery that can safely claim to have no superior, perhaps not even a rival, in recorded human history.

Our history books are generally too busy to deal with such phenomena. Time was when historical records dealt with human heroes rather than with human movements. But the heroes held up to popular admiration were of other mould than Newton. Warriors who had slain, or caused to be slain, their thousands and their ten thousands, kings and emperors who used peoples and lands as pawns in their game of power and empire, royal favourites and jousting nobles—all these and their myrmidons strutted across the page of history; and if in the mob represented as gazing with admiration and envy at the doings of their masters, could be discovered a Roger Bacon, a Shakespeare, or a Newton, warming themselves in the patronage of the great, or struggling with the prejudices and bigotry of their rulers, this was sufficient condescension and digression from the main object of the narrative.

Things are indeed different now. The outstanding personalities of history are made to fit into a truer picture of general human progress. Popular movements, national aspirations, industrial development and social and political ideology now claim the attention of the historian. Yet it is only natural that men and women who stood somewhat aside from general life, and whose activities did not immediately affect national fortunes, should figure but modestly in the story.

It is nevertheless a fact that while King and Parliament were struggling for supremacy, while Cromwell was with dry piety and powder pursuing the fugitive remnants of Stuart royalism, while rehabilitated royalty was revenging itself on its oppressors, there was emerging a power in the intellectual field of human activity whose supremacy has been almost unchallenged for centuries, and bids fair to remain unchallenged for generations to come.

We all love tales of adventure and stories of derring do. But need adventure be conceived as restricted to the sphere of physical strife? When man had to hew out for himself a precarious habitation in the primeval forest, in the face of inhospitable nature and the rivalry of man and beast, the facing of physical danger and the unflinching acceptance of personal hazard were the supreme virtue, the essential concomitant of human survival. Now that the races of humanity are still doomed to struggle with one another for the possession of the earth, and each race, nation or tribe conceives it its first duty to defend its claim to the free and undisturbed enjoyment of the fruits of its labours, there is virtue in physical prowess and popular applause for the heroes of national offence and defence. May we not hope for a better day when adventure will not connote battle, suffering and death, when human conquest will have for its aim the widening of the human empire over the forces of nature, when applause and admiration will be the reward of intrepid inquiry into the physical universe in which we live and thoughtful examination of the moral universe that lives within us?

The heroes of this type of adventure, the warriors to whom fall the spoils of such conquest, are at least as worthy of our admiration as king and emperor. Founders of religious systems and movements, prophets and moral teachers of humanity, philosophers and scientists, poets and masters of great fiction, have moulded human life at least as efficaciously as politicians and favourites. There is as much adventure in the exercise of the intellectual and spiritual faculties as in the exercise of the physical faculties, and there is as much romance in the story of these adventures as in the narratives of knightly tournaments and martial glory.

Isaac Newton was born on December 25, 1642.*

* *Note on the Calendar* :—The Gregorian calendar was introduced in 1582, but was not adopted in Great Britain till 1752. All dates given in this book are the contemporary English dates, i.e. in accordance with the Julian, or Old Style, calendar. One exception will be made. Till 1752 the English new year was on March 25th, with the result that, e.g., what was called in England January 1, 1710, was January 12, 1711 in the Gregorian calendar! It would

THE LINCOLNSHIRE CHILD, 1642–1661

His parentage was undistinguished and of decent obscurity—not poor but by no means wealthy. The father—also Isaac Newton—belonged to a family that had farmed the modest Woolsthorpe estate for several generations. The mother—Hannah Ayscough—came from Market-Overton in Rutlandshire. Later in life Newton himself had ideas of relationship to some Scotch Newtons of noble descent. But this nobility was not apparent at his birth. Isaac Newton's father died only a few months after the marriage, and several months before the birth of his only and posthumous child; the effects on the physical health of the mother were such that Newton was born prematurely, diminutive in size and sickly in health.

Woolsthorpe Manor House, in which Newton was born, is still in an excellent state of preservation, inhabited by people whose family has resided there for several generations. The room in which the birth took place is still used as a bedroom, and a tablet over the mantelpiece records the birth of Newton with the couplet:

> Nature and Nature's laws lay hid in night,
> God said " Let Newton be," and all was light.

The house is pleasantly situated in a very fertile valley, a few minutes from the Parish Church of Colsterworth, where the entry of Isaac Newton's baptism on January 1, 1643, is still to be seen.

Isaac lived in this house for many years. At first he was taken care of by his widowed mother, who also had a small estate at Sewstern. But a neighbouring clergyman, Rev. Barnabas Smith of North Witham,

be too confusing to adhere to this practice here, and so our dates will be in accordance with the Julian calendar, but with the new year at January 1st, so that English January 1, 1710 corresponds to Gregorian January 12, 1710. The difference between the English and the Gregorian calendars in Newton's lifetime was ten days till February 28, 1700, and eleven days after this date.

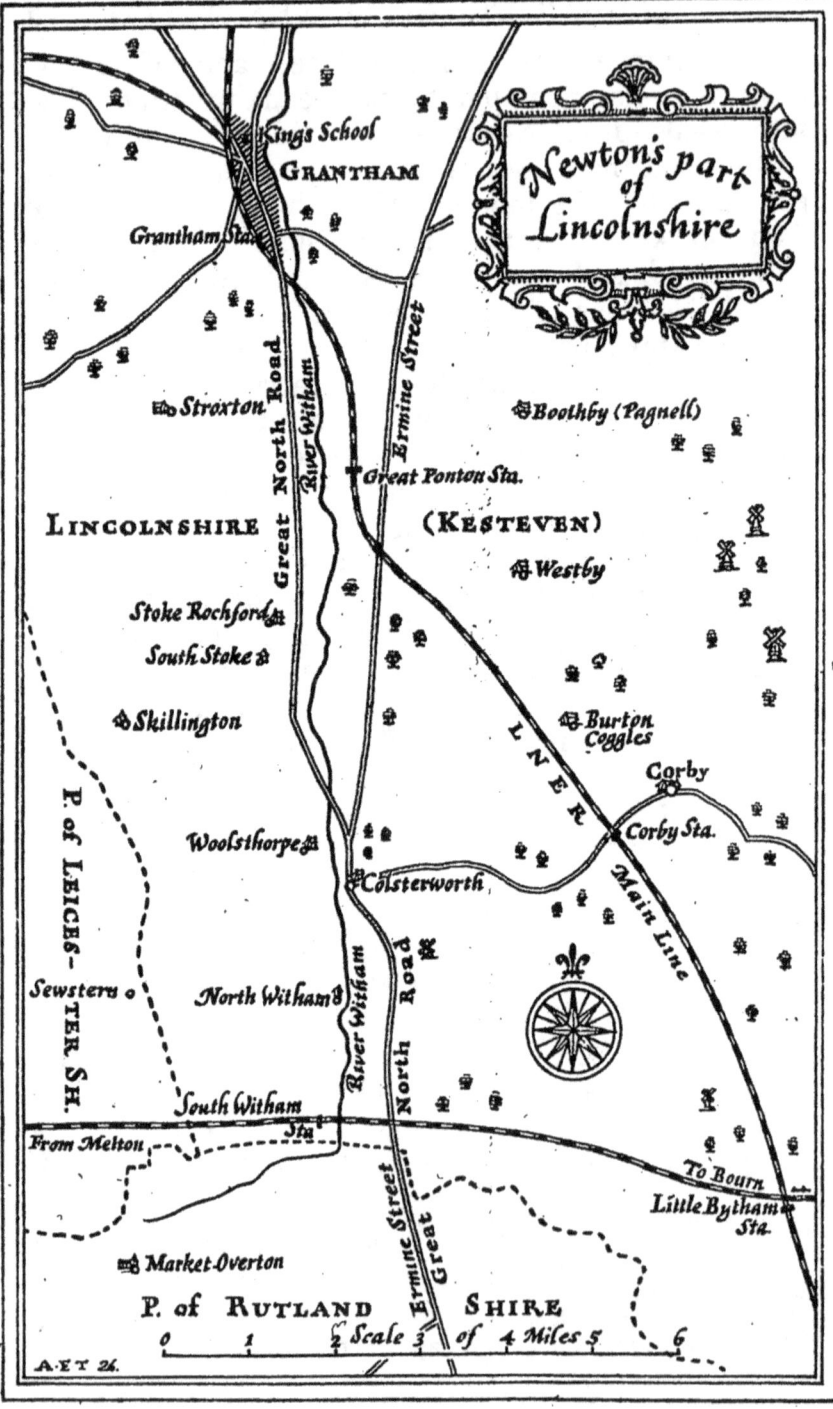

THE LINCOLNSHIRE CHILD, 1642-1661

an old bachelor whom his friends persuaded to marry at last, married Newton's mother in 1645. Young Isaac's upbringing was then entrusted to his maternal grandmother; he remained at Woolsthorpe while his mother was with his stepfather in North Witham. He went to small schools at Stoke and Skillington, and seems to have enjoyed the care and encouragement of his maternal uncle, Rev. W. Ayscough of Burton Coggles. Newton's paternal relations were apparently but little interested in him till he became famous and could be of use to them. A paternal uncle lived in Colsterworth, and it was his grandson, John Newton, who, as heir-at-law, inherited the estates Newton had obtained from his father and mother at Woolsthorpe and at Sewstern. This relative was no great ornament either to Newton or to humanity. He sold the estates to Edmund Turnor of Stoke Rochford, whose lineal descendants have been the owners of Newton's birthplace ever since.

When Isaac reached the age of twelve he had exhausted the very meagre intellectual resources of the villages near Woolsthorpe, and in 1655 he was sent to the King's School, Grantham. He was not expected to walk the six miles to Grantham every day, and so he lodged at the house of an apothecary and his wife, Mr. and Mrs. Clark, next to the George Hotel in the High Street. The house has disappeared, and its site is now occupied by an extension of the George Hotel. King's School, Grantham, still exists in a modernized form, a modest building in pleasant surroundings. The Old School, in the form of a small hall, where the teaching took place in Newton's time, has been well preserved. Its dimensions are $75\frac{3}{4}$ feet by 27 feet, it has a stage at one end, and is now used for school assemblies only. The walls and beams bear witness to the proclivities of schoolboys in all generations to cut their names wherever possible. Little did

I. NEWTON, who carved his name on one of the beams and his initials on another, realize that he was to carve his name in an indelible manner on the walls of the temple of fame.

The head master's house is still used as such, but we need conjure up no romantic scenes of youthful genius discovered by an admiring and careful mentor. Isaac Newton did not at first attract any attention at all. He seemed an ordinary boy, perhaps not very strong, and not given to taking any pains to excel in anything. We have Newton's own authority for stating that he paid little attention to his studies, while at play he had not much in common with his schoolfellows. There was nothing of the infant prodigy about young Isaac: he possessed dormant powers that needed awakening.

It seems absurd to have to say that the awakening took the form of a kick in the stomach. A boy who ranked higher than Newton in the academic hierarchy of the school once caused him considerable pain in this way. Newton retaliated in lusty manner, fought his assailant till he would fight no more, and duly rubbed his nose against the wall of the Grantham Parish Church—which stands by the school—to emphasize the victory of revenge.

Newton was not satisfied with this revenge. His enemy stood above him at school, and so Newton was determined to vanquish him here too. He finally succeeded, and the impulse of the effort carried Newton on till he reached the proud position of top boy of the school.

For Newton possessed remarkable reserves of mental power, and his apparent idleness had been the result of his preoccupation with matters in which he was more deeply interested. He was low in school at first, but when stung to action he soon outstripped all competitors. He did not play games with the other

THE LINCOLNSHIRE CHILD, 1642-1661

boys; but he did better than play games—he invented new ones. He flew and experimented with kites. He was silent and thoughtful in disposition, but naughty enough to attach paper lanterns to the kites and frighten the simple peasants with the belief that they saw those dreadful heavenly visitors—comets. He was remarkably good with his hands and could manipulate tools with great skill, constructing various mechanical contrivances, including a windmill model, which was made independent of the winds of heaven by means of the power supplied by an imprisoned mouse. Young Newton was an excellent observer of the motions of the heavenly bodies : he made sundials which were used for a considerable time—one still exists in the Parish Church at Colsterworth. He drew well and even wrote some indifferent verse. Newton was, in a word, a lad of singular ability, doing well anything that came his way.

As we have seen, Newton was not averse to a pugilistic encounter when the necessity arose for one, but in essence he was gentle and not given to rough play. He seems to have preferred the company of girl friends to that of boys, and was ever ready to render them any service of which he was capable.

In 1656 Newton's stepfather died. His mother, Mrs. Smith, came back to Woolsthorpe, bringing with her the three children that had been born to her and the Rev. Barnabas Smith. They were two girls and a boy : Mary, Benjamin and Hannah Smith. Isaac must have contracted a great friendship with his stepbrother and two stepsisters : all through his life he helped them, and he made them his heirs to a very considerable personal property. A niece, the daughter of Hannah, became a favourite of Newton's later in life, and she kept house for him when he lived in London.

Newton had reached the top of the school. He was

as well educated as one could expect to be in those rural surroundings. His mother needed a man to take charge of the family estates, and what more natural than to summon Isaac home and invest him with the authority and the responsibilities of this post —especially as he was the lawful heir to the Woolsthorpe manor and had been given the Sewstern estate when his mother married Mr. Smith? The young Newton, then in his fifteenth year, came back from Grantham. No doubt, if he had been interested in farming and grazing, in buying and selling, Newton would have made a success of his modest patrimony. But the truth is that his interest never wandered in the direction of agricultural pursuits. His work on and in connexion with the farm was of no real value, and it soon became clear that what he hungered for was opportunity for academic study and mechanical manipulation.

In September, 1658, Oliver Cromwell died. The heavens sent a violent storm to signalize this event. Instead of seeing to it that the force of the gale should do no damage to the buildings of the farm, to the produce and the cattle, Isaac spent his time in jumping with the gale and against the gale—hoping in this way to estimate the force produced by the wind. We can conceive with what haste Newton's relatives decided that farming was not the vocation to which he was born, and how pleased they were to send him back to the Grantham school and his boy friends, to the apothecary's house and his girl friends.

There is nothing that can be said to characterize scientific genius as regards matrimony; some of the world's greatest mathematicians and astronomers remained unmarried, others had matrimonial experiences both happy and unhappy, while still others have not been wanting in the spirit of adventure in this field. Copernicus was a priestly bachelor. Tycho

THE LINCOLNSHIRE CHILD, 1642–1661

Brahe ran off with the proverbial village maiden and led a married life of alternate storm and calm. Kepler married twice : he treated the matter of marriage in a spirit of business-like caution and achieved an imperfect bliss nevertheless. Newton is credited with two adventures, both very simple and innocent in character.

The apothecary's second wife had a daughter by her first husband, Miss Storey by name. She was Newton's junior and playfellow, good-looking and clever. Apparently Newton's pleasure in the company of girl friends developed into something more mature in the case of Miss Storey. But both he and his lady friend were not rich enough and not poor enough to do anything improvident. Newton gave up any ideas he may have had of married life, and devoted himself to his studies. He finally went off to Trinity College, Cambridge, on the advice of his uncle from Burton Coggles ; Miss Storey stayed in Grantham. Years later Newton returned to find her married to another (and to yet another !). To his credit be it said that Newton retained the memory of his early and tender regard for Miss Storey during a long life, and extended to her an ever-ready hand of friendship and help.

Generosity was one of the outstanding features of Newton's character all through his life, and he was always making gifts or loans to all who needed such aid. In a notebook that Newton kept in 1659 we have ample evidence of his readiness to help where help was needed. We learn that he spent some money on " chessemen and dial," while Newton also places on record the fact that he spent some money on " china ale," " bottled beere," " sherbet and reaskes," and other desiderata of the mortal part of his make-up.

From 1658 to 1661 Newton was at the Grantham school preparing for Cambridge. His local reputation grew quickly, and when he left school his departure was made the occasion of a laudatory address by the head master. On June 5, 1661, Isaac Newton, then in his nineteenth year, was admitted to Trinity College, Cambridge, and embarked upon a career of complete devotion to science.

CHAPTER II

THE CAMBRIDGE STUDENT, 1661–1665

> And soon the flimsy sums he floored,
> And conned the conics thro' and thro'.
> S. K. Cowan:
> "A Supplemental Examination"

IT has been suggested that the time of reaction and repose after violent political convulsion is usually characterized by unparalleled progress in science, literature and art. The Napoleonic wars were followed by great developments in all branches of science in Britain, France and Germany, and the recent European war has certainly left behind it an interest and an activity in science which is producing very fast scientific progress in all civilized nations. Whether we have here a relationship of cause and effect, whether humanity in its best minds, being thoroughly ashamed of its recent bestiality, strives to make amends by more devout application to the matters of the mind, we cannot say. It is certainly of interest that the religious convulsions in the England of the middle of the sixteenth century were followed by the Elizabethan period of drama and literature, while the political disturbances of 1642–1660 were followed by one of the most brilliant periods of British science, the establishment of the Royal Society and the foundation of the Royal Observatory at Greenwich.

When Isaac Newton left Woolsthorpe for Cambridge his local reputation did not imply that he was really a very learned youth. He knew comparatively little, especially in mathematics. Other students had done more than he. This is, in fact, one of the greatest advantages of going to a university like Cambridge. He who in his modest home-circle is acclaimed as the wise and learned man and excites the admiration of his relatives and neighbours, has to submit to comparison with, and the competition of, the best intellects brought together from all corners of the land. Woolsthorpe, Colsterworth and Grantham could be proud of their Isaac, but Cambridge was quite unmoved by Newton's arrival. Other students were at least as far advanced in their studies as he, and young Newton did not immediately carry off all the prizes within his ken.

This was no real disadvantage to Newton. On the contrary, Newton derived considerable benefit from his comparatively late development. It is very seldom that the youthful prodigy carries out the promise of his early achievements. It is not often that the boy or girl who is urged on prematurely to advanced studies reaches real knowledge and understanding of the fundamental ideas involved in these studies. Nature does not approve of being hurried. Peaceful bodily development accompanied by calm mental growth, with no hot-house and artificial forcing, yield the best results. The rural surroundings of Newton's early years, far from the incitement of intense competition and the rush for cheap reputations, enabled him to reach healthy mental and physical maturity, untired by premature exertions.

It was an additional advantage to Newton that he did not go to the University till well into his nineteenth year of age. The modern aspect of university life, which is mainly an avenue to the professions, coupled with the poverty of most of the students, tends to send young boys and girls to the universities when they should still be at school. The university is a place of *independent* study within the reach of scholars of outstanding merit and authority. For independent study an essential desideratum is maturity, culture and a sense of responsibility. Newton was not hurried: he came to Cambridge just when his intellect was arriving at full development.

The vigour of his intellect was soon displayed to his tutors. On more than one occasion he was found to have mastered independently treatises that were to form the topics of extended lecture courses. The result was that he was released from attending such courses. He could roam at will in the domains of knowledge, and he was soon attracted to the studies with which his name was to become immortally associated.

Kepler's "Optics" is the first book that Newton read at Trinity, and as we shall see it was in optics that he made some of his first and epoch-making discoveries. Attracted by a book on astrology which he picked up at a fair held near Cambridge, Newton realized his ignorance of geometry. He therefore bought a copy of Euclid's "Elements." It is indeed remarkable that he was not led to Euclid in a more direct manner, and, further, that he did not realize the genius of the writer and the importance of the subject. To his very practical mind, not given to idle speculation and reaching conclusions by a marvellous insight and intuition, Euclid's efforts to prove what appeared to be self-evident truths suggested that the author was merely trifling—a mistake made so often and so disastrously in our own generation. Newton suffered for this mistake. He entered Trinity as a sizar, a class of student then, and till about a century ago, required to perform various menial services in return for his tuition and necessaries of life, or "commons" as they are still called—for Newton's mother was not rich and he had to use the force of his own intellect to raise him in the academic scale. When he competed for a scholarship in April, 1664, the examiners, while recommending him for an award, commented on his poor knowledge of geometry, and Newton was thus led to reconsider his opinion of Euclid.

Newton's interests were indeed hardly those of a mathematician as such. His mind was essentially and primarily that of an experimenter. His school days were spent in a manner that suggested the doer rather than the thinker, and in the apothecary's house he acquired a leaning to alchemy and chemical experimenting. His main incentive to study was his desire to understand natural phenomena — be they the motions of the planets and comets, the periodic flow of the tides, the beautiful colours in the spy-glass and in soap bubbles, the resistance of the air and the laws of motion, the properties of substances and the transmutations of the metals. "Observe the products of nature in several places, especially in mines . . . and if you meet with any transmutations out of their own species into another (as out of iron into copper, out of any metall into quicksilver, out of one salt into another, or into an insipid body, etc.), these, above all, will be worth your noting, being the most luciferous, and many times lucriferous experiments too in philosophy." So wrote Newton to a friend a few years after his undergraduate days—long before he had forsaken Cambridge for the London Mint, and gold and silver had acquired a personal professional interest for him.

Cambridge was not then the home of English mathematics, which flourished rather in Oxford and London. It was Newton's own genius that gave to Cambridge its mathematical renown. He studied the algebra of his time, and the new methods of René Descartes, and at once began to have ideas wherein lay the germs of the binomial theorem and of the differential calculus.

The school boy (or girl) of to-day is taught so well that one often wonders if he can learn at all. Good teaching does not consist in reducing all the hard nuts to an amorphous powder, and all the hard crusts to soft pap, which can be swallowed without effort, and which escape assimilation into the system owing to the lack of the saliva of mental effort. The learner must learn and not be merely taught; the effort must be his at least as much as the teacher's. When mathematical teaching is a careful cataloguing of all possible questions that might be invented by evil-minded examiners, and literary study has become a careful cataloguing of insignificant events in the authors' lives, coupled with pathological investigations into peculiarities of accidence, syntax and misprints of successive editions, who will wonder that grown-up Englishmen do not rush to see Shakespeare's dramas and do not feel the romance of the binomial theorem?

Yet the binomial theorem is full of romance; it throbs with human interest. It is an adventure into the unknown. Lead the learner on by judicious encouragement to generalize: $(1 + x)^2 = 1 + 2x + x^2$, $(1 + x)^3 = 1 + 3x + 3x^2 + x^3$, etc., into $(1 + x)^n = 1 + nx + \frac{n(n-1)}{1.2} x^2 + \ldots + nx^{n-1} + x^n$, where x is any number and n is any positive whole number. He cannot but see in this a conquest of territory visible to the eye, not mysterious or clothed in the shadows of night, but the necessary consolidation of empire, the appropriation of lands without which present possessions are only loosely held.

But should not the young learner palpitate with excitement when urged to venture into uncharted depths and take n to be absolutely anything? The series of terms:

$$1 + nx + \frac{n(n-1)}{1.2}x^2 + \frac{n(n-1)(n-2)}{1.2.3}x^3 + \ldots$$

has no end if n is not a positive whole number: it goes on and on and on, indefinitely, to infinity. Does this mean anything? Can, for instance, the square root of $1 + x$ be as much as a whole and unlimited series of terms, all following one another in logical and inexorable law, and taking us—who knows whither?

We are too civilized in our modern mathematical teaching. The ruggednesses are smoothed out, and the polished surface thus produced leads to smooth and effortless gliding—perhaps oftener to involuntary skidding. He will be a benefactor of the race who will contrive to bring back into our schools a sense of hazard and adventure, the spirit that characterized those who made and accumulated the knowledge that is taught there. Then we shall no longer suffer the boor who claims with pride that he never could understand mathematics, that lessons on algebra and geometry were occasions for inattentiveness, and that the binomial theorem is to him merely the topic of ignorant jest.

A great discoverer of our own time has said that the truly great discoveries are made in early manhood, that he who has passed beyond his thirtieth year without having contributed such discovery to the sum of human knowledge, is doomed to go to the grave without making any great discovery at all. If true, this thought is a sad one. It is, at any rate, remarkable that many of the world's most epoch-making discoveries fell to the lot of youth and not of old age. In the case of Isaac Newton, two grand discoveries which have rendered his name pre-eminent among all men—the calculus, and universal gravitation—were made in germ long before he reached his thirtieth year, nay, before he was twenty-four years of age; while he discovered the spectrum and the true nature of colour before he was twenty-eight.

The binomial theorem, if not so important as his other discoveries, would yet have ensured Newton a permanent niche in the temple of fame even if he had done nothing else, and this theorem he constructed early in 1665, at about the same time as he took the degree of Bachelor of Arts, when just over twenty-two years old!

Newton's reputation at Trinity was not, however, commensurate with his powers and with the things he was doing in his study. We have already seen that he did not emerge from the scholarship examination of 1664 without some censure on the score of inadequate geometrical knowledge. His general behaviour was not out of the ordinary. He was quite human in his pursuits and tastes. He visited the beerhouse, played cards for money—and lost, and when he went home for vacations he brought as presents for his mother and stepbrother and stepsisters the same kind of childish gifts that the Cambridge man of to-day carries to doting relatives.

We have no information about the order of merit of the graduates of 1665. Was Newton the outstanding graduate of his year? Was he, perhaps, quite undistinguished in this examination, without any obvious signs of his future—in fact, immediately emerging—greatness? We cannot tell. We do know that he was not elected Fellow at once. He was busy with the binomial theorem, he invented the method of fluxions—his form of the differential calculus—he thought about colours and was being led to experiment with a prism, within a few months of his becoming B.A. Yet he kept all these things to himself. A few notes are extant written in his own hand and containing references to these and other matters. But Newton was never inclined to making public any unfinished or half-baked discoveries. By the time that people round him began to suspect that he was doing things of value, he was well on the way to becoming the great discoverer and original thinker that he was soon recognized to be.

CHAPTER III

THE DAWN OF MATHEMATICAL DISCOVERY, 1665

" Line upon line, line upon line ; here a little and there a little."
Isaiah xxviii, 13

THE Great Plague broke out in 1665 and spread from London to Cambridge. This put an end to Newton's regular studies at Trinity, for the college authorities decided not to face the risk of the pest overtaking them with all the students in residence. They therefore dismissed the college on August 8, 1665. This went on for practically the whole of the time till December of the same year, for the whole of the second half of 1666, and for some time in 1667. Newton left college in 1665, even before the dismissal in August. These two years were thus a time of intermittent labour. But we shall soon see that they were by no means wasted.

We have records in Newton's own hand which date his pure mathematical progress with considerable accuracy. Referring to the winter between 1664 and 1665, i.e. about the time of his taking the Bachelor's degree, Newton says : " At such time I found the method of Infinite Series ; and in summer, 1665, being forced from Cambridge by the plague, I computed the area of the Hyperbola at Boothby, in Lincolnshire, to two and fifty figures by the same method." In a notebook of Newton's we find an account of his first discovery of fluxions, written on May 20, 1665, and another account with applications dated November 13, 1665. Now we know that Newton was away from college during the twelve weeks ending December 21, 1665. It is therefore clear that he made excellent use of his enforced exile from Cambridge.

Modern mathematics is unthinkable without the calculus. In order to study any natural phenomenon, it is essential to deal with the rates of change of various quantities. In heat we want to deal with the rate of change of temperature, or the rate of heating or cooling. In dynamics we have to deal with rate of change of position, or the velocity, with rate of change of velocity, or the acceleration. In engineering practice we have to deal with the rate at which work is being done. In pure mathematics the shape of a curve is defined by the rate of change of the direction of the tangent, while the tangent itself indicates the way in which position on the curve changes in terms of any two quantities or co-ordinates that define such position.

The differential calculus is the branch of mathematics that deals with such topics. When once the suggestion of the fundamental significance of rates has been made it becomes clear that progress along this direction is an essential condition of progress in science at all. Newton realized this, and in his theory of fluxions, i.e. the rates of flowing or changing quantities, he discovered the first practical step in the development of the calculus.

A geometrical form of the problem is the following: Given a curve drawn on a sheet of paper, how can we find its area or the value of any area defined by a piece of the curve and straight lines? This is of fundamental importance, for it is soon seen that if we plot velocity against time on squared paper, the area between the curve thus obtained, the time axis, and two ordinates parallel to the velocity axis, gives the

THE DAWN OF DISCOVERY, 1665

distance described. The engineer in his indicator diagram plots the pressure in the cylinder of his engine against the position of the piston: the area of the closed curve obtained during a backward and forward stroke of the piston gives the work put into the machinery by the steam or gases in the cylinder.

How can such areas be measured? We can count squares and by making the squares small enough we can, with sufficient labour, obtain useful results. But such a method is only approximate and empirical—the mathematician would desire, of course, an exact solution by calculation. What is the exact area of a circle, of an ellipse, of a piece of a parabola cut off by any chord, of the space between an hyperbola, an asymptote and two lines parallel to the other asymptote? How can they be calculated theoretically?

The Greeks had faced this problem and applied to it their incomparable genius. The method of exhaustions was invented for the purpose, and in the hands of a master like Archimedes yielded useful results. The area bounded by a closed curve may be considered to lie between the areas of two polygons, one inscribed and the other circumscribed to the curve. The greater the number of sides in the polygons the closer do their areas approach one another, and the more accurately therefore can the area of the curve itself be calculated. Let the numbers of sides be increased indefinitely. The areas of the two polygons approach one another indefinitely, giving ultimately the exact area of the curve itself.

Thus, take a circle. Divide its circumference into, say, four equal parts at A, B, C, D. Draw the square ABCD. Draw the tangents at A, B, C, D: they form another square A'B'C'D'. Then the area of the circle lies between the areas of the two squares. Taking the radius to be unity, we at once get that the area of a circle of unit radius lies between 2 and 4. Now divide the circumference into eight equal parts at A, B, C, D, E, F, G, H : we get a regular octagon ABCDEFGH inscribed to the circle, while tangents at these eight points define another regular octagon, A'B'C'D'E'F'G'H', which surrounds the circle. The area of the circle lies between the areas of these two octagons, i.e. between 2·828 and 3·314. Taking sixteen points we get that the area lies between 3·062

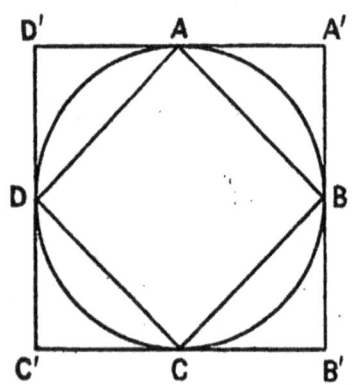
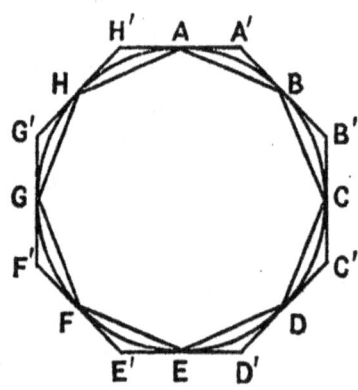

FIG. 1
AREA OF CIRCLE

Squares inscribed and circumscribed to circle.   Octagons inscribed and circumscribed to circle.

and 3·182. It will be seen how the areas of the polygon approach one another, and it will be clear how, proceeding in this way, we can, by means of pure calculation, reach any degree of accuracy required.

The area sought for is what mathematicians call π. By means of the method of exhaustions Archimedes showed that π lies between $3\frac{1}{7}$ and $3\frac{10}{71}$.

The area of a piece of a parabola cut off by a chord was also obtained—and in this case easily and exactly —by a process based on the same idea. Yet, clever though the idea may be, and great the mental vigour of the Greeks, only a few cases could be dealt

THE DAWN OF DISCOVERY, 1665

with in this way—a really practical process did not exist.

An extension of the area problem is that of finding the volume enclosed by a surface. The German, John Kepler, of whose work we shall hear more in the sequel, was a man of remarkable mathematical genius. Being in possession of unexpected money one day, he decided to lay down a stock of wine. The wine was to be delivered in casks, and naturally the question arose as to the volume that is contained by a cask. This volume depends on the outline of the cask, for a cask is approximately defined by the revolution of a curve round an axis of symmetry, and Kepler set himself the problem of calculating the volume in terms of the curve which defines the shape.

Kepler had some, but not complete, success. Whether he ultimately purchased the wine, and whether its qualities satisfied his expectations, need not concern us now. This incident and Kepler's failure served to stimulate the Italian Cavalieri to put forward in 1635 the epoch-making suggestion, that an infinite number of points make a curve, an infinite number of curves make a surface, an infinite number of surfaces make a volume. This is not correct logically, and should read : a curve consists of an infinite number of infinitely short straight lines, a surface consists of an infinite number of infinitely narrow strips, a volume consists of an infinite number of infinitely thin sheets. But Cavalieri's suggestion is one of the most important in the history of mathematics, and many useful applications of his idea were soon obtained, especially by the Frenchman Fermat, some of whose wonderful theorems are still puzzling the greatest mathematicians of to-day.

No general rule was discovered, however, till the Englishman, John Wallis, of Oxford, found one for dealing with a whole class of areas, namely, such as are defined by curves in which one co-ordinate is proportional to a power, like the square or cube, of the other co-ordinate. So long as this power was a positive whole number no difficulty existed: but when the power was negative or fractional Wallis had no means of obtaining the exact result.

Thus, suppose one co-ordinate varies inversely as the other, as in the curve given when Robert Boyle plotted the volume against the pressure for a perfect gas: this curve is called a hyperbola, and this is the problem referred to by Newton as having engaged his attention when he fled to Boothby, in Lincolnshire, because of the plague. The young Newton, then twenty-two years of age, found the complete law for dealing with any function involving roots of complicated expressions, by means of the binomial theorem.

But Newton went further. He realized that Cavalieri's ingenious idea contained the germ of something far greater. If a curve consists of an infinite number of infinitely short lines, cannot the length of the curve be considered as growing by the process of adding on little line after little line? If a surface consists of an infinite number of infinitely narrow strips, cannot the area of the surface be considered as growing by the process of adding on narrow strip after narrow strip? If a volume consists of an infinite number of infinitely thin sheets, cannot the volume be considered as growing by the process of adding on thin sheet after thin sheet? Consider a length, area or volume, not as something fixed and dead, cut and dried, but as something alive and developing, something that varies continuously, as flowing water varies from moment to moment: look at it cinematographically and not by means of an instantaneous and rigid image on a dead screen.

This is an idea that has changed the face of

mathematics. The conception of physical quantities, including geometrical position, length, etc., as fluent, the study of rates of flow and of change, opened the way to unlimited development. The problem of finding the direction of the tangent at any point of a plane curve, is at once solved by asking : What is the ratio of the flows of the two co-ordinates that define the curve, considered as growing quantities ? In other words, we differentiate. To find the area of a curve we have to ask the inverse question : What is the quantity whose rate of flow divided by the rate of flow of one co-ordinate gives the other co-ordinate ? In other words, we integrate. The youth of twenty-two had not only solved one of the problems of the ages : he could give an impulse to mathematics leading to centuries of novel progress and growth. The differential and integral calculus, the theory of differential equations, and the vast ramifications of applications and developments, might have been traced to the genius of the barely graduated scholar of Trinity.

If Newton's discovery did not immediately set free the developments that we can now see should have resulted therefrom, if the discoverer had to face the possibility of being denied the credit of his own originality, he had himself to blame. Be it out of modesty, or out of a sense of extreme cautiousness, Newton kept the discovery to himself, and denied to the world of science what was its obvious due, namely, a published statement of his discovery.

Newton wrote brief accounts of the method of fluxions as early as May and November, 1665. He developed it further in a manuscript dated May 16, 1666, and still further in a tract which he wrote in October, 1666. In the last he made twelve applications of the method, to curvature, concavity and convexity, maxima and minima, etc., many of the applications being new and of importance. But the plague raged on, Newton worked on and invented the conception of second fluxions, or what we now call repeated differentiation, and even his closest mathematical friends knew nothing of the great events taking place in this obscure youth's brain.

CHAPTER IV

THE GERM OF UNIVERSAL GRAVITATION, 1666

> Sir Isaac Newton was the boy
> That climbed the apple tree, sir;
> He then fell down and broke his crown,
> And lost his gravity, sir.
>
> J. A. SIDEY: "The Irish Schoolmaster"

THE Great Plague and Newton's enforced absence from Cambridge meant a serious break, or succession of breaks, in his work. His attention had already been directed towards optical problems, and he was preparing to undertake suitable experiments for the purpose of elucidating some of the puzzling things about light and colour, when he had to give up the work and retire into the country. He made up for this by special devotion to the pure mathematical studies: yet Newton's mind was essentially practical and experimental, and subconsciously searched for natural phenomena and their explanations.

Enforced idleness is not always a curse—to the deep thinker, such mental leisure and freedom from immediate preoccupation with some definite task, are a positive advantage. The man of science, indeed, often performs his function most effectively just when, to the unsuspecting observer, he is enjoying a mental holiday. The reception of impressions and their undisturbed incorporation in the mental equipment of the scientific worker finally give rise to the idea that, like the prince's kiss, removes the charm that hinders progress, and substitutes for it a vigorous vitality.

Removed from his college rooms with their preparations for optical experiments, deprived of the solace of mathematical treatises which only a college library could supply, Newton was forced to spend much of his exile in quiet observation of things round him, in meditation on problems that occupied the attention of his generation. The most insistent problem, one which demanded a solution and which was obviously ripe for the application to it of some outstanding intellect, was that of gravitation.

When man emerged from the childlike state in which things were taken to be what they seemed, he was forced to the conclusion that the earth upon which he lived was not the bottom of a sort of pie-dish, with the sky as a cover. Little by little it was realized that the earth must be disconnected all round, and in the last few centuries B.C., and in the first few centuries A.D., the spherical shape of the earth became the accepted view of all thinkers. The heavenly bodies, consisting of the sun and moon and the small specks of light loosely called stars, were seen to move. The sun and moon were found to move from east to west across the sky each day, but the path described varied from day to day during the year and month respectively, in a more or less regular and periodic manner. Of the stars, which all appeared to move across the sky from east to west each night, all, except a few, moved regularly in exactly the same paths night after night, although the times of rising and setting of any particular star varied regularly from night to night. It was soon observed that this corresponded with the motion of the sun. The resulting view adopted was therefore that the heavens as a whole rotate round the earth from east to west each " day," this day being called sidereal or stellar. The sun and moon were

UNIVERSAL GRAVITATION, 1666

assigned additional motions round the earth from west to east, the sun going round once in a year, the moon once in a month.

But there were exceptional stars, to which the name planets was given. These were unlike the sun and moon in that they appeared on the sky as tiny specks, but they imitated the sun and moon in that they had additional motions round the earth. These planets were Mercury, Venus, Mars, Jupiter and Saturn. The last three, namely, Mars, Jupiter and Saturn, went round the earth from west to east on the whole in periods of about 2, 12 and 30 years respectively: but complications existed in the form of annual loopings, the planet stopping in its west to east motion, then actually moving back, i.e. east to west, then stopping again and resuming its west to east motion. Mercury and Venus executed motions that were at first not at all clear to visualize. They seemed to hug the sun in a peculiar manner, appearing sometimes to the east of the sun as evening stars, then becoming invisible and reappearing on the west of the sun as morning stars, and so on indefinitely. It was finally accepted that Mercury and Venus also go round the earth, but unlike the other or outer planets, they both take a year to go round the earth, and execute loops in other and shorter periods, 88 days per period in the case of Mercury, 225 days per period in the case of Venus.

There thus arose the famous Ptolemaic system of the universe in which the earth was taken as spherical and absolutely fixed at the centre of the universe, the whole universe rotating round the earth once a day, and the moon, Mercury, Venus, the sun, Mars, Jupiter and Saturn going round the earth once each month, year, year, year, 2 years, 12 years, 30 years respectively, with no loop in the moon's motion, an 88-day loop in Mercury's motion, a 225-day loop in the motion of Venus, no loop in the sun's motion, and annual loops in the motions of Mars, Jupiter and Saturn. The circuit and loop were in each case represented as motion on a deferent and epicycle, the planet moving on the circumference of the epicycle whose centre moved on the circumference of the deferent.

The Ptolemaic system, first given to the world in the second century A.D., soon became accepted doctrine. When European nations sank into semi-barbarism—at least, as far as the arts and sciences were concerned—in the early Middle Ages, the Arabs and the Jews kept the Ptolemaic astronomy alive, just as they kept other products of Greek science alive, and even made some significant additions. Such additions were essential since, after all, the only test that can be applied to a scientific theory is comparison with fact. If in political controversy facts are not seldom modified so as to make them fit more comfortably into some particular theory, it is nevertheless true that in science facts are supreme. A theory is indeed a summing up of a number of facts. If I say that two volumes of hydrogen explode with one volume of oxygen to produce two volumes of water vapour at the same temperature and pressure, then I do not mean that some tribunal has decreed that hydrogen and oxygen must do this and refuse at their peril. What I mean is that whenever the experiment has been tried it has been found to yield this result. We therefore, with that faith in the reasonableness of nature which we call causality, assume—perhaps with pathetic if unconscious humour in the eyes of a superior intelligence—that the same result obtains in all cases, even when not observed by us, and will obtain in future. This is a theory about water. Let but one case arise in which this result is not obtained, and the theory begins to wobble, and unless the deviation from theory is adequately accounted for, the theory must give way to a more adequate one.

Ptolemy's theory was not only qualitative, it had

UNIVERSAL GRAVITATION, 1666

to be quantitative too. With their marvellous sense for philosophic values the ancient Greeks realized that a theory must be quantitative, that it is not sufficient to assert vaguely that the moon goes round the earth, but we must state exactly what the path is and how fast the motion is at any moment. Now the Greeks knew practically nothing about the science of dynamics (the other contemporary nations knew next to nothing of anything). There are several reasons why this should have been the case, as we shall soon see. The Greeks therefore had to make some basic assumptions, which we now know to have been without justification. They divided matter into earthly and celestial. Earthly matter was subjected to the earth's influence and moved up and down, perpendicular to the earth's surface. Heavenly matter, on the other hand, was free of this materialistic influence, and so moved in a perfect manner, i.e. on circles round the earth with constant speeds along the circles.

To our modern ears such statements sound bizarre. Why should uniform motion on the circumference of a circle be the perfection of motion? Why should some matter be called celestial and granted the privilege of this perfect motion? The foundations of dynamics have been changed since then, and Newton had a great share in producing the change. Yet it must be remembered that even to-day there is in at least some of us a lurking belief that the heavenly bodies are somehow heavenly, while matter on our earth is after all only earthly. People still exist who dub as materialistic the astronomy that considers all matter in the universe as of the same kind as the matter of and on our earth, in spite of the fact that the unique position of the earth in the economy of the universe is no longer believed in even by such muddled intellects.

According to the Ptolemıc system then, all the heavenly bodies moved on circles with uniform motions. The so-called fixed stars called for no further comment. In the case of the planets Mercury, Venus, Mars, Jupiter and Saturn, deferents and epicycles had to be introduced, but the centre of the epicycle moved uniformly round the deferent and the planet moved uniformly on the epicycle. For the sun and moon no epicycles seemed necessary.

But the facts began to be troublesome. When the position of a planet at any moment was calculated from this theory and compared with the observed position in the sky, the agreement was often bad, in fact very bad, and this applied to the sun and moon too. Modifications had to be introduced. The centre of the sun's annual path was taken to be different from the centre of the earth, and when this did not suffice the uniform motion of the sun on its circle was taken with reference to angular motion as seen from some third point. And so on: in the case of each body such and similar complications were introduced. And all through the Middle Ages these complications accumulated. The instinct that underlay this process was excellent—fact is supreme and theory must be modified to suit the facts. But clearly something was wrong somewhere. For there is another instinct in us which sees beauty and truth in simplicity, and when a scientific king scoffed at the astronomers by saying that had he been present at creation he could have given some good advice, he was merely giving expression to the healthy desire of science to arrive at simple truth which needs no frills and furbelows to make her plausible.

Views other than Ptolemy's had been expressed in ancient times, and suggestions had been made that possibly things are quite different. Must we assume the earth to be fixed, and at the centre of the universe?

UNIVERSAL GRAVITATION, 1666

Perhaps the earth rotates. Perhaps the earth's centre moves. Perhaps the earth is a body moving round another body, and perhaps this other body is the sun. Such views were, however, effectively silenced, particularly during the later Middle Ages, when Ptolemaic astronomy finally conquered Europe, and became invested with supreme authority. Authority was the keynote of those days. Truth existed somewhere, and the search for truth meant the search for hidden meanings in the pronouncements of authority, be it in the Bible, in Aristotle, or in any other accepted and unquestionable source.

A spherical earth was indeed sufficient departure from primitive and naïve belief. To envisage humanity as standing all round a globe with heads pointing outwards in all directions already did violence to a psychology that saw a physical and intellectual anchorage in a flat and stably-fixed earth. To deny the fixity of the earth and to make it wander about in space while gyrating on its axis meant deranging still further the mental equilibrium of humanity.

New ideas were not welcomed in those days. The initiation of a scientific revolution nowadays meets with wide approval—revolution is often met with open arms for its own sake, and accorded a publicity worthy of better causes. But in the later Middle Ages scientific originality brought with it a publicity of an undesirable kind and notoriety of an unpleasant character. We can therefore appreciate all the more the intellectual compulsion that forced Nicolas Copernicus to discard at least a part of the Ptolemaic system.

Everything pointed in the direction of change. If the sun is one of the planets going round the earth, why should the epicycle periods in the case of Mars, Jupiter and Saturn, and the deferent periods in the case of Mercury and Venus, be exactly the same as the period of the sun's motion round the earth, namely, a year? Why are the sun and moon exceptional in not having loops in their paths? The planets are very far away from us: the nearest to us, the moon, has a distance from us sixty times as great as the earth's radius (it is true that in Copernicus' time the moon's distance was taken to be somewhat less, but this does not diminish the force of the argument); the sun is still further from us, the planets are still more distant, and the fixed stars are even further off. Does all this universe (and Copernicus did not, in fact, have any adequate idea of how vast the universe really is) revolve round the earth, and as fast as once in a day of twenty-four hours?

We cannot believe direct and untutored visual impressions. We must use our brains and reason out sensible conclusions. Things are not what they seem. Our impressions of motion are relative: the man on the cliff sees the ship depart from the shore, while the man on board the ship sees the shore and cliff depart from him. Copernicus was forced to the grand conclusion that the earth is not fixed and is not at the centre of the universe. The earth rotates on her axis once a day: this explains in rational manner the apparent rotation of the whole universe round the earth every day. The earth revolves round the sun once a year: this explains in rational manner the loops in the paths of Mars, Jupiter and Saturn, which also go round the sun in circles greater than the earth's orbit, and the peculiar hugging of the sun by Mercury and Venus, which go round the sun in circles smaller than the earth's orbit. The marvel of the exact year period of the epicyclic motion in the case of the outer planets, and the exact year period of the deferent motion in the case of the inner planets, now disappears.

A beautifully simple solar system emerges, slim in conception and unencumbered by deferent and

UNIVERSAL GRAVITATION, 1666

epicycle. The sun is exceptional and has no looping motion, because it is indeed the exceptional member of the system, the monarch reigning at its centre, the hub of the universe : the moon has no looping motion, because it is also exceptional—it is not a planet at all. The moon is a satellite, attached and subservient to the earth, going round the earth and playing only a secondary rôle in the solar system—the Copernican system.

Public attention was drawn to this view only after the death of Copernicus in 1543. A howl of indignation went up from established authority, but Copernicus was fortunate in being dead, for all that could be done to him was to execrate his name and forbid the reading of his book and the dissemination of his system. It is generally believed that the Catholic church was mainly responsible for this. It is not so : all constituted authority, Catholic and Protestant, religious and secular, opposed the new astronomy. The opposition, in fact, was the slothful indignation of human intellectual inertia, that refused to be moved from its moorings, and that feared to give up the fixed earth and wander about with bearings lost in a bewildering universe. Why, even the very week was in danger : for were not the seven days of the week, sanctified by Biblical command, associated with the seven planets of pre-Ptolemaic astronomy ?

Yet Copernicus triumphed. After Galileo's discovery of the satellites of Jupiter, the phases of Venus, the discs of Mars, Jupiter and Saturn, no other view than that of Copernicus could survive. By the middle of the seventeenth century no enlightened person who was free from papal and other backward religious authority could do other than accept the Copernican system.

It is the irony of human progress that the revolutionary very soon finds himself outdistanced by his disciples. Copernicus' system was a tremendous advance on the old system, but his dynamical knowledge was hardly, if at all, in advance of Ptolemy's. In abolishing the central position and fixity of the earth, Copernicus paved the way for the abolition of Greek ideas of motion based upon earthly and celestial types of matter, and caused the emergence of the problem of dynamics—the why and the wherefore of motion and changes of motion—with peculiar insistence. But he could not liberate himself from the perfect, i.e. uniform circular, motions of the planets, including the earth, round the sun, and of the moon round the earth. In Copernicus' great work, "De Revolutionibus Orbium Coelestium" (1543), epicycles and excentrics are still piled upon one another: one marvels, in fact, at Copernicus' genius all the more in that he did manage to convert the earth into a rotating planet revolving round the sun, while he was himself yet subjected to the bondage of Greek dynamics, and had to use epicycles for other purposes.

John Kepler was the man who actually liberated himself and mankind from the effects of Greek dynamical ideas in relation to astronomy; he took the step that was necessary to complete Copernicus' work. Basing himself on the careful observations of Mars made by his teacher, Tycho Brahe—the greatest instrumentalist and observer of the Middle Ages—and also by himself, he discovered the true paths of the planets.

Kepler realized at the outset of his work that circular motions had to go. He searched for the true paths of the planets. He had no clue except the rather doubtful one that somehow the ancient Greeks had, for purely theoretical reasons, studied with considerable care the curves known as conic sections, the curves obtained when a circular cone is cut by a plane. This study had been part of mathematical education even since,

UNIVERSAL GRAVITATION, 1666

and Kepler thought of trying whether the paths of Mars and the earth round the sun could be taken to be conics and give results in accordance with Tycho Brahe's observations. The conics had to be closed, i.e. ellipses. It was a sheer guess, the leap forward of the man of genius.

If the earth and Mars go round the sun in ellipses, what is the relation of each ellipse to the sun? An ellipse has three pre-eminent points inside it—the centre and the two foci: the latter are the points used for drawing an ellipse quickly, since, if a string lies on a board with its end-points fixed, and is stretched by means of a pencil which runs round it, then the pencil describes an ellipse of which the end-points of the string are the two foci. We may guess that if the path of a planet is an ellipse, then either the centre of the ellipse or one of its foci will be at the centre of the sun. Which is the correct point?

If we have the correct point, the next question is: How does the planet move round the elliptic path? What is its speed at any point of the path? If Kepler had possessed any true dynamical knowledge then he would have found out the law of this speed very easily—he might have forestalled Newton. But dynamical progress came too late for him. He had to discover the truth by sheer heavy arithmetical calculation. It was a mighty task that would have been beyond the powers of any other man. Kepler held out. His calculations were to him the romantic quest for a celestial gift—he pursued them with the ardour and devotion of a fanatic. He laboured for a decade, without even the help of the logarithmic tables that were then being invented and calculated by Napier and Briggs in Britain. He tried hypothesis after hypothesis—finally he triumphed. He discovered the laws that go by his name—Kepler's laws of planetary motion (1609):

I. Every planet moves on an ellipse with a focus at the centre of the sun;

II. The line joining the centre of the sun to the centre of the planet sweeps out equal areas in equal times.

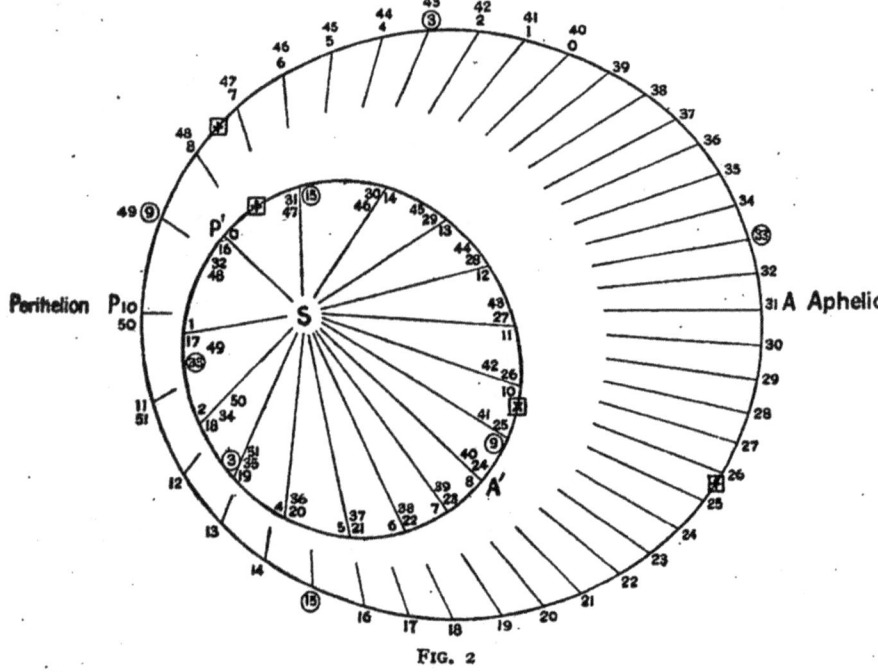

FIG. 2

KEPLER'S LAWS

Simultaneous positions of two planets in their orbits. AP/A'P' = 1·84, so that ratio of periodic times is $1·84^{3/2} = 2.5$. Circles represent oppositions of the two planets. Squares represent conjunctions of the two planets.

The first law is intelligible, but what does the second law mean? Kepler did not know. He was the discoverer of the truth but not of the significance of this truth. It was a great achievement to discover the true paths of the planets. It was the coping-stone to the Copernican system, the completion of the revolution initiated by Copernicus, which did away once and for all with the deferents and epicycles, excentrics and equants of mediaeval astronomy. Kepler put into the Copernican system the spirit of modernity. How Copernicus would have rejoiced to learn the truth from his posthumous disciple!

Kepler had yet another problem to solve—the problem of his youth. Was there any relationship between the distances of the planets from the sun? He had foolish speculations at first, but he finally hit upon a wonderful and simple truth. His third law (1619):

III. The square of the time taken by a planet to describe its path completely once round the sun is proportional to the cube of its mean distance from the sun

is indeed a peculiar law, savouring of the mysterious that Kepler loved so well.

A new problem now began to emerge. Whereas the question so far had been: *How* do the planets move? the question now became: *Why* do the planets move in the manner discovered by Copernicus and Kepler? The motions of the planets had been found: they had to be explained. This became particularly urgent when Kepler published his laws. One might perhaps ascribe uniform circular motion to celestial matter and flatter oneself that one is doing something sensible— but it becomes difficult to do this conscientiously with elliptic motion round a focus in which the "radius vector" sweeps out areas proportional to the time. Planets may be deities and as such choose to march sedately and majestically in perfect circles with unvarying speeds—impervious to all influences tending to hurry or to delay them. But it is difficult to envisage a planetary deity at the end of a sort of municipal rolling broom, that obeys trade-union rules and sweeps out so much area per day—no more and no less.

Kepler himself tried to explain the motions that he had unravelled. He failed. A sort of gravitational influence of the sun on the planets was suggested by him, even the law of the inverse square of the distance suggested itself to him and to others in France, in Italy and in England—but to no effect. A dynamical explanation of planetary motions was out of the question so long as the foundations of dynamics had not been discovered.

It is for this reason that the Vortex Theory of Descartes enjoyed such a vogue. To explain with considerable plausibility the whirling of the planets round the sun, and of the satellites round the planets, one need only observe the way in which whirlpools in water carry with them the bodies caught by them, forming minor whirlpools or eddies, which exert local influences of the same kind. Descartes therefore postulated vortices in an ocean of " ether," filling up the whole of the space occupied by the solar system.

Quantitatively, Descartes' theory breaks down, but it held the field because there was no competitor. Vague ideas about gravitation could not dethrone a theory supported by the authority of this great philosopher and mathematician. Views based upon sound dynamical principles alone could rout the vortex theory.

Fortunately, the dynamical principles were soon available. Galileo's great astronomical discoveries have had the effect of eclipsing somewhat the lustre of his fame in the history of dynamics. If astronomy suffered from the arrogance of self-constituted authority, it had at least one advantage—observation of the skies was encouraged both by nature and by man. A man is indeed no higher than the beasts of the field if he is not attracted by the aspect of the heavens, especially

UNIVERSAL GRAVITATION, 1666

in the clear skies of Italy. The observation of the heavens was also encouraged (from mixed motives) by kings and princes. When man believed that the heavenly bodies shone for his benefit and so as to enable him to foretell the future of his petty affairs, detailed study of these bodies was an essential constituent of government and statesmanship. The emperor Rudolf may have meant astrology when he installed Tycho Brahe in Prag—Tycho Brahe could get on with his astronomy nevertheless. Astronomical discovery was thus in a sense inevitable. In the case of dynamics, however, if nature did occasionally suggest feebly: "Observe the motions of bodies and the causes of these motions," man did not respond to the invitation. Why observe the motions of bodies when Aristotle has told us all about them already? This would only tend to encourage scepticism and to undermine authority.

And dynamical laws required very exceptional powers of observation and of intuition for their unravelling. It is difficult nowadays to measure accurately the velocity with which a body is moving at any moment—in the early seventeenth century it was quite impossible. It was even difficult to observe accurately the position of a falling body from second to second, in an age when the measurement of time did not yet enjoy the help of modern clocks and chronographs. Further, motions are liable to all kinds of disturbances—air resistance, friction, surface irregularities, and so on. Finally, where was the mathematical machinery for dealing with motions and changes of motion before Newton invented it in his method of fluxions?

It was Galileo who ventured to dispute the authority of Aristotle in the dynamical field. He experimented and he discovered that the rate of fall of a body near the earth does not depend upon the weight of the body—if this weight be sufficient in relation to the size of the body to make the effect of air-resistance negligible. Bodies falling from the same level reach the ground after the same interval of time. Finally he discovered that a falling body moves in a remarkable manner—its speed increases with the time in a perfectly regular manner, it is accelerated uniformly.

The regular increase in the speed of a falling body is of course due to the pull of the earth on the body, what we call the body's weight. It must follow therefore that a steady pull produces a steady *increase* of velocity—a new and a revolutionary idea. To the forerunners, and even to the contemporaries of Galileo, motion connoted force—wherever motion existed force had to exist to explain it. In Galileo's mind a new idea began to dawn : not motion connotes force, but change of motion connotes force. A body moving under no force enjoys constant speed in constant direction : let a force act upon it and its speed or its direction or both will vary. Galileo could not prove this directly, since a body moving under no force cannot be realized. But the man of genius sees further than the limited range of the available experience—he makes the leap that carries the human soul forward on the wings of discovery, and Galileo was a man of exceptional genius.

Here was the beginning of true dynamics. The ancient belief was—no force, no motion. The modern view is—no force, no change of motion, i.e. no acceleration. Need we wonder at the circumstance that this truth lay so long undiscovered ? Think of the ox-cart in which Homer's father went to market—think of the hardships of locomotion in those days, and right down nearly to our own generation. The only smooth road known then was the sea—and the sea gave the seafarers of those days sufficient to think about to divert

UNIVERSAL GRAVITATION, 1666

their minds from dynamical speculations. It needed a mighty intellectual effort of scientific abstraction to dissociate locomotion from the jolting and the sense of effort that accompanied it, and to see in the ideal smooth and unvarying motion the raw matter, so to speak, of dynamics.

The discovery of the true laws of motion in 1637 set the road free for progress in the dynamics of astronomy. Men of the greatest eminence sought for the solution of the problem of the planets. Robert Hooke in England—a man of considerable, if erratic and unstable, genius, the discoverer of the *ut tensio sic vis* of elasticity—speculated on this problem, and even announced that no doubt he could solve it by combining the conception of gravitation with the dynamics of Galileo—if only he could hit upon the correct law of force to attribute to the sun's gravitation, the exact way in which this force depended upon the distance of a planet from the sun. It was the youthful Newton, meditating in the exile of his Lincolnshire garden, who approached the problem with a direct frontal attack, remarkable and historic even in its initial failure.

Tradition relates that one day in 1666 Newton was sitting in his garden at Woolsthorpe, when he saw an apple fall from a tree : this suggested gravitation to him, that the fall of the apple was due to the earth's pull, and hence arose his discovery of the law of gravitation. There is a rival tradition which makes Newton observe the fall of the apple through a window on the first floor of the manor house. The apple tree itself became the object of general interest and historic significance, and it was kept standing till blown down by a gale in 1820. A chair made from the wood of this tree is to be seen at Stoke Rochford. Descendants of the tree are said to be thriving to this day; visitors to Woolsthorpe, who come as pilgrims to the house where Newton was born, do not fail to inspect the descendant that is still to be seen in the garden there. Popular imagination clings to this apple incident—yet it is clear that Newton was not the first man to see an apple fall, or to ask himself why an apple falls. Newton did not discover that the earth pulls the apple—the earth's gravitation was common knowledge in ancient times, and Galileo had, when Newton was still unborn, discovered all about the fall of an apple or of any other body free to give way to the earth's pull. The fall of the apple—if it was really responsible for initiating in Newton's mind the train of thought that culminated in the discovery of the law of gravitation—did it in the following manner, if one may venture to put into feeble words the flash of Newton's genius:

"Why do the planets go round the sun? Why do they not move in straight lines? Evidently there is a force pulling them out of the straight-line path at every moment, and clearly this force is due to the sun. The moon goes round the earth and does not go off in a straight line. This must be due to the earth. Ah! An apple has just fallen to the ground: the earth has pulled it down. How far up does the earth's influence extend? We know that no matter how high up we go—to the summits of the highest mountains—this influence exists without obvious weakening. Does the earth's gravitation extend to any distance, no matter how great—perhaps even as far as the moon? Can this be the force that compels the moon to accompany the earth, to travel round and round the earth indefinitely as the earth travels round the sun?

"Yes, this is a pretty theory: can it be proved? Can it be shown that the pull required to explain the moon's motion is just that afforded by the earth's gravitation? Any attempt at such proof must postulate some law according to which the gravitative pull of the earth varies with the distance from the earth;

for clearly we cannot suppose this pull to be the same for all distances, even to the ends of the universe. It must diminish as the distance increases. What is the law of this diminution? Suppose that one body is twice as far from the earth's centre as another: what is the relationship between the pulls exerted by the earth on these bodies?"

The most important step in the solution of any natural problem is the formulation of the problem itself. Newton formulated his problem in masterly manner—it only remained to attack the solution. Two steps had to be taken. The first was to determine how a gravitational pull would vary with distance from the gravitating body. The second was to see whether the earth's pull on the falling apple could also account for the motion of the moon.

How should a gravitational pull vary with distance? A very simple consideration at once suggests a solution. Suppose that from a point O there is a gravitational influence in all directions. Take two spheres with centres at O, one having double the radius of the other. The same gravitational influence is spread out over the surfaces of the two spheres: but the sphere of double radius has a surface area four times that of the smaller sphere. It follows that the gravitational pull at a point on the larger, double, sphere must be one-quarter of the pull at a point on the smaller sphere. This is at once generalized into the inverse square law, namely, that the pull exerted at any point is inversely proportional to the square of the distance of this point from the gravitating body.

Such considerations would run through Newton's mind; but Newton was remarkably realistic in his outlook upon nature and her problems. An ounce of observational or experimental proof was worth to him more than a pound of theoretical argumentation. This characteristic accompanied him all through his long career of scientific activity. Newton looked for observational justification of the inverse square law—and the third law of Kepler, discovered nearly half a century before, and quite ununderstood all that time, gave him the proof he desired.

The planets move on ellipses: but these ellipses are nearly circles—this was indeed why circular motions were insisted upon till Kepler's success in emancipating us from them. Except for the planet Mercury, the planetary paths are so nearly circles that, for a rough calculation, we can ignore their ellipticities and assume them to be circles with their centres at the centre of the sun. Further, the motions on the paths are variable —yet the variations are really very small and we can for a first approximation imagine the planets to describe circular paths with constant speeds.

It needed the mental courage of a Newton to thrust himself back in this way into the primitive views of Greek astronomy—uniform circular motions. Newton was not yet mathematically ripe to solve the problem of the planets in their fulness. With characteristic clarity of vision he pounced upon the essential and discarded the inessential. He felt that what he needed could be obtained by such approximate argumentation.

Now if a body moves on a circle with constant speed there must be a force producing the continuous change in direction of the motion; this follows from Galileo's dynamical discoveries. Every schooboy who has done a little mechanics knows that the force required is the mass of the moving body multiplied by the product of the speed into the rate at which the direction of motion is rotating. This means that for radius r and angular velocity n the force per unit mass of the moving body must be the speed rn into the rate at which the direction of motion rotates, namely, n. Hence the force per unit mass must be $n^2 r$.

But Kepler's third law states that the squares of times taken by the planets to travel once round the sun vary as the cubes of their mean distances from the sun. If then T is the time for a planet moving in a circle of radius r we get T^2 proportional to r^3. But the time of a complete revolution varies inversely as the angular velocity n, since the time multiplied by

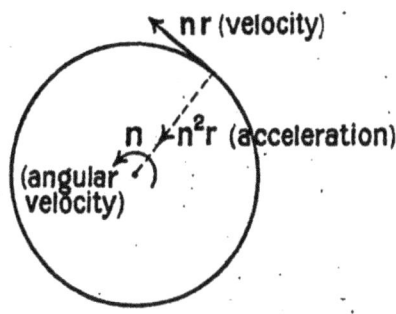

FIG. 3
CENTRAL ACCELERATION IN CIRCLE

If body moves with constant angular velocity n in circle of radius r, acceleration is n^2r towards the centre.

the angular velocity must give the complete angle all round, or four right angles, i.e. the same for all the planets. Hence n^2 is proportional to $1/r^3$ so that n^2r is proportional to $1/r^2$. This is of course the inverse square law!

Newton then argued by analogy that if the gravitational pull of the sun on the planets follows the inverse square law, then the pull of the earth on the apple and on the moon must also follow the inverse square law. Why? Because Newton had a profound belief in the unity of the universe, in the similarity of effects produced by similar causes—that truth is independent not only of time but also of place, so long as at different times or at different places we look for the results produced by exactly similar causes. If the motion of the moon is to be explained by the earth's gravitation then it can only happen on the basis of the inverse square law.

The distance of the moon from the earth's centre is about sixty times the radius of the earth itself. It follows therefore that the gravitational effect of the earth on a unit mass of the moon must be one part in 3,600 of the effect produced on a unit of mass near the earth's surface, say, a falling apple. The accelerative effect on the moon must therefore be 1/3,600 of the accelerative effect on the apple. The accelerative effect on the apple is to add a speed of 32·2 feet per second each second. Hence the accelerative effect on the moon must be a speed of 32·2/3,600 feet per second added each second, i.e. 0·00895 feet per second added each second.

Is this accelerative effect equal to n^2r, where r is the radius of the moon's path in feet and n is the angular velocity of the moon round the earth in true mathematical measure, i.e. in radians* per second? We know r to be $60R$ where R is the earth's radius in feet. As regards n we know that the moon describes 360° round the earth in about 27⅓ days: this makes n equal to 1/375,000, with the result that the acceleration produced by the earth's pull on the moon is $R/2,344,000,000$ where R is the earth's radius in feet.

Everything depends upon R, the radius of the earth. This is a very difficult quantity to measure, but Newton lived at a time when accurate knowledge of the size of the earth had already been obtained. Yet he was far removed from books and he had to rely on memory. It was commonly supposed that the size of the earth is such that a degree of latitude is 60 English miles. Newton adopted this view and made the complete

---

* A radian is the angle subtended at the centre of a circle by an arc of the circumference of length equal to one radius.

UNIVERSAL GRAVITATION, 1666

circumference 360 times this, or 21,600 miles. The radius was therefore 3,440 miles or about 18,160,000 feet. The acceleration of the moon should thus be 0·00775 feet per second added each second.

Galileo gives 0·00775, Kepler gives 0·00895. The latter exceeds the former by nearly sixteen per cent!

"This is not equality," said Newton to himself. "My thought has been but an idle speculation. This great problem is not to be solved by me. Yes, I can patch up my theory and imagine all kinds of causes to account for the apparent deficiency in the moon's angular motion. But I can only deal with realities, with forces which I know something about. To postulate mysterious agencies whose effects I cannot measure directly is not a useful contribution to the problem."

We must pause and observe this young man, twenty-three years of age. By sheer genius he has hit upon the very secret of the heavens, and carried out calculations in which lie the germ of the complete explanation of the motions of the planets. Yet Newton does not dare to conclude that his theory is right, because there is a discrepancy of sixteen per cent in the quantitative investigation. The leap of genius must not be confounded with intellectual recklessness. The discoverer must be true to himself. His ego is indeed an invaluable ingredient of his work—but it must be the ego of honest personality, not that of selfish egotism. We can picture the pang with which Newton abandoned his theory, how his innate intellectual honesty struggled with human weakness in his closet on the first floor of the Woolsthorpe manor house—the tiny room that can still be visited and admired as the hallowed shrine of the true spirit of discovery—and conquered. If ever the scientific method scored a victory it was in the humble home of the widow and her orphaned children.

Not a word did Newton say of all this to a living soul. He had failed and there the matter ended. It did not really end there, for about sixteen years later Newton discovered the source of the discrepancy and the cause of his failure—the simple fact that the value of R he had assumed was too small, and that he should have taken it by about fifteen per cent greater! Perhaps humanity is all the better for the apparent failure and the delay in the ultimate success. Is not this example of intellectual honesty and of impersonal devotion to the cause of scientific truth as valuable as any contribution to human spiritual values ever made by mortal man?

CHAPTER V

THE ANALYSIS OF LIGHT AND COLOUR, 1666

> Behold, a prism within his hands
> Absorbed in thought great Newton stands, . . .
> The chambers of the sun explored,
> Where tints of thousand hues were stored.

THE Great Plague was responsible for interrupting an experimental investigation which turned out to be Newton's main work for a decade. The discovery of the telescope and its remarkable use at the hands of Galileo naturally made optics in general, and telescopes in particular, the subject of keen interest on the part of all the scientific men of the seventeenth century. Newton's attention was drawn to the subject in a very natural manner. He himself records that early in 1664, when he was still an undergraduate, he observed and measured lunar crowns and halos. What Newton did in optical research during the next two years we cannot say, but we do know that in the beginning of 1666 he bought a prism to examine the phemonena of colours. He tells us himself that having discovered why the telescopes of those days were defective, he was led to consider the making of telescopes based upon reflection. "Amidst these thoughts I was forced from Cambridge by the Intervening Plague, and it was more than two years before I proceeded further." Newton did not in fact proceed to construct a reflecting telescope till late in 1668. It is therefore clear that the first notions of Newton's great discovery of the spectrum must have come to him in 1666. We know that he was absent from Cambridge in 1666 up to about the end of March, and then from about the beginning of July onwards till the end of the year ; hence it is highly probable that Newton's first glimpses into the true nature of colour were obtained early in 1666, and that he made some experiments between April and June of that year. Newton was then twenty-three and a half years of age.

A telescope is based upon the following fundamental idea. An image of a distant object is formed by means of a lens—called the object glass—and this image is then observed by means of another lens—called the eye-piece. The construction of the most complicated modern telescopes is based upon essentially the same idea. It therefore seems to be an indispensable condition of a good telescope to make the object glass collect the rays from any point of the object observed accurately to a focus. If rays from any point of the object do not give an exact point image, then the telescope suffers from aberration and its optical value is diminished—the picture in the telescope is indistinct, which is bad for the eyesight and bad for exact measurement.

It is, however, a remarkable fact that the only form of optical operation that brings rays emanating from a point to an accurate and exact focus at another point is reflection at a plane surface. The perfectly plane mirror gives a reflection which is perfect in sharpness of outline and of detail. Any other optical operation, especially refraction or light bending at a plane or any other surface between one medium, say air, and another, say glass, gives more or less blurring and indistinctness.

This fact was soon understood and the problem of optics narrowed down. If it is impossible to get an exact image of all points of the object, it may yet be possible to secure an exact image of *one* point

LIGHT AND COLOUR, 1666

of the object—for scientific, especially astronomical, purposes this would be a great thing, nearly as much as is really needed for useful work.

Even this modest demand is not easily satisfied in an ordinary telescope arrangement, in fact is quite

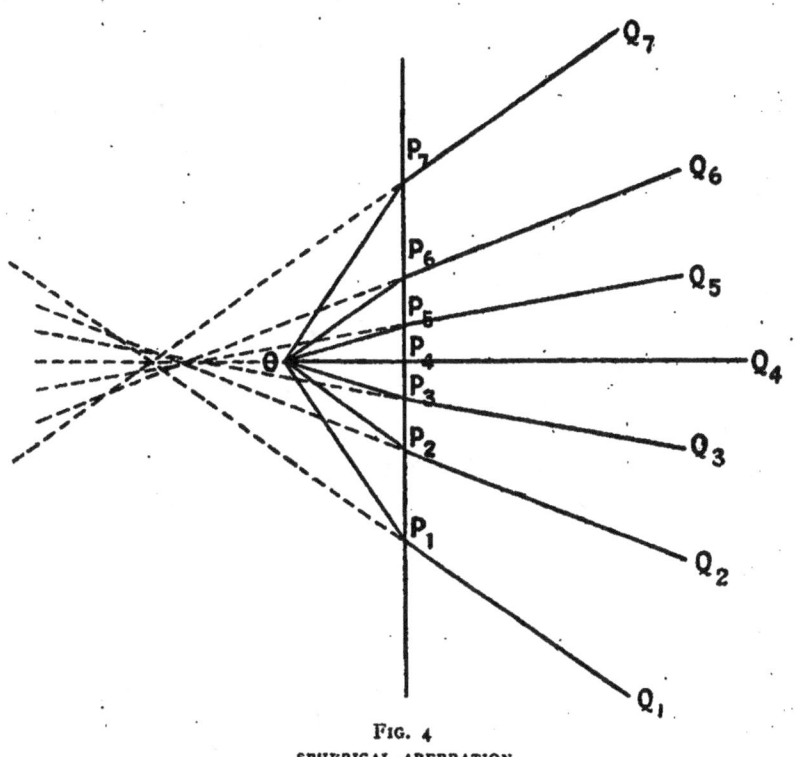

FIG. 4
SPHERICAL ABERRATION

Rays OP_1, OP_2, OP_3, etc., from the point O, are bent into the directions P_1Q_1, P_2Q_2, P_3Q_3, etc., when passing from air into glass; these new directions are not concurrent; hence "spherical aberration."

out of the question if the surfaces of the object lens are plane or spherical. Thus if $P_1P_2P_3$ . . . represent a plane surface of separation between air and glass, and OP_1, OP_2, OP_3, . . . are rays from a point O striking the surface at P_1, P_2, P_3 . . ., then the new rays after refraction, namely, P_1Q_1, P_2Q_2, P_3Q_3 ..., do not pass through one point—they form in fact a very complicated bundle of rays which do not give a sharp image point at all. The same applies to refraction—as this is called—at a spherical surface of separation between air and glass. How is this difficulty to be overcome?

This "spherical aberration" engaged the attention of the mathematicians of the generation immediately preceding Newton. When Galileo first invented and used his telescopes he added to astronomical science an instrument of marvellous power—but the telescopes he used were nevertheless very defective instruments, and no doubt Galileo's blindness towards the end of his life can be traced to the strain he imposed on his eyesight in using these telescopes so assiduously. The cause was taken to be the spherical aberration just indicated, and search was made for a means of eliminating it.

It was the great Descartes who suggested what appeared to be a really good way out of the difficulty. He looked at the problem in a somewhat different way. Instead of assuming that lenses must have plane or spherical surfaces he asked himself the following question: "Given an object point O, rays from which are to be refracted at some surface of separation between air and glass, so as to pass through another and image point I, what should be the form of surface used?" Only in one case can a spherical surface be used, and this is of no value for telescopic purposes—it is really of fundamental importance for high-power microscopic objectives as constructed nowadays. Descartes found that to get the surface he wanted he had to depart from the spherical shape, and grind the lenses in accordance with new curves that bear his name—Cartesian ovals: he therefore suggested the use of these new surfaces, and even devoted time and

LIGHT AND COLOUR, 1666

energy to the devising of machines for producing such surfaces on glass.

Yet it was all a fallacy. In actual fact the grinding and polishing of non-spherical surfaces on glass is on the whole impracticable, and spherical surfaces are the basis of applied optics to this day. But even if perfectly ground and polished surfaces of exactly the form indicated by Descartes' theory could have been produced, they would have yielded practically no improvement in the optical properties of telescopes. It was not spherical aberration that was to blame for the imperfection of Galileo's telescopes. The blame belonged really to a cause of which Galileo, Kepler and Descartes were all ignorant—an ignorance all the more striking since so many natural phenomena flow from the same source.

The colours of the rainbow, with its promise of peace and reconciliation after the storm, the flash of the diamond, the colours of the prism, the charm of the soap-bubble, why the very things observed in the telescopes of the period, seemed to beg for elucidation by an almost obvious principle—yet the principle lay hidden from the human understanding, and needed the genius of Newton to give it to the world. Already in 1666, when twenty-three years of age, Newton had decided to experiment with the prism, and had reached the momentous and fundamental conclusion that the telescopes of Galileo and Kepler were incapable of improvement, and that a new principle of telescopic construction had to be employed.

It seems to be impossible to say how much knowledge of the phenomena of colour Newton had acquired when he reached this conclusion. He did not publish an account of his researches till five or six years later, and in this account a clear chronology is not given. We find moreover that when about three years later—in 1669—he helped his great teacher and friend Isaac

Barrow in the publication of a set of lectures on optics, Newton did not even then correct Barrow's quite absurd notions about colour. We must therefore conclude that Newton himself was not quite clear on the subject, even after he had abandoned attempts to improve Galileo's telescope. The fact, however, that Newton did reach the conclusion to discard Descartes' idea and to turn to reflecting telescopes instead, cannot but indicate that he had discovered the main principle underlying such conclusion.

The fact is of course that in addition to the spherical aberration difficulty of which the great mathematicians before Newton were quite aware, there exists another cause of indistinctness and imperfection in refraction which they did not suspect, and which Newton discovered. We need not devote much space to the description of this discovery and the method adopted by Newton to make it. Every man or woman who claims to possess the rudiments of scientific culture surely knows that when a beam of parallel white light is sent through a prism so that it is deflected by refraction, it does not emerge as a beam of parallel white light, but rather as a bundle of beams of all colours—no longer parallel but pointing in different directions. Newton sent a beam of sunlight of circular section through a prism, so that it was bent once on entry into the prism and once on emergence from the prism: when caught upon a screen two effects were observed. The screen gave a coloured patch of light instead of a white patch: this in itself was not so very unexpected, since Newton, like millions of intelligent human beings before him, had observed that prisms produce colours—this is why prisms are so popular for purposes of adornment round lights and in windows. But the second effect observed was unexpected and thought-provoking. The original beam was circular in section: the patch of coloured light on the screen was lengthened in a direction at right angles to the edge of the prism; and it was no moderate lengthening—the patch was five times as long as it was broad. This was completely at variance with all contemporary doctrine on the passage of rays through prisms.

Perhaps Newton did not at once solve the mystery —judiciously chosen and carefully executed experi-

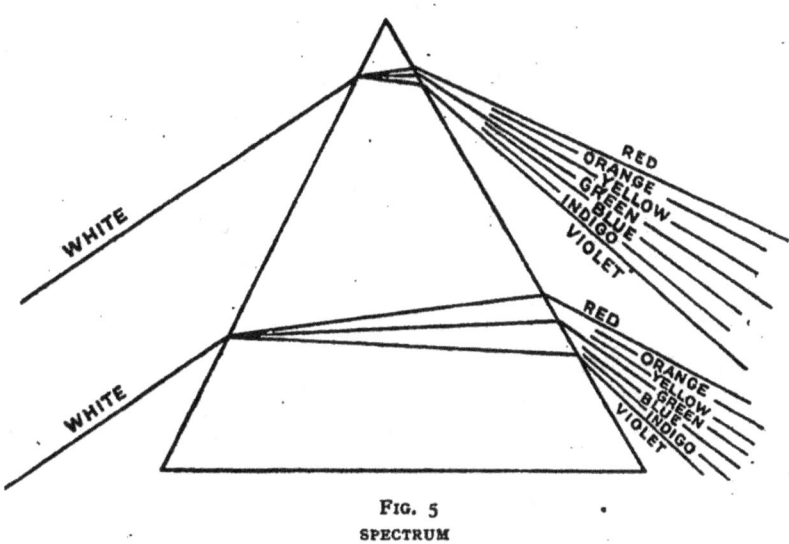

Fig. 5
SPECTRUM

White light split up into coloured rays when passing through a prism, forming the spectrum.

ments were yet necessary in order to explain this new phenomenon, and these Newton carried out during the succeeding five years. But one thing was obvious to him. If refraction through a prism converts a thin circular beam of light into a beam whose section is a long strip, then clearly a telescope that depends upon refraction through a lens cannot but produce imperfect, coloured and blurred images. This did not depend upon the exact shapes of the lens surfaces. Cartesian ovals were simply irrelevant to the problem. A lens cannot produce a clear and sharp white image of a star—colour must be introduced and cause confusion: the aberration to be eliminated was not spherical aberration but "chromatic" aberration, error due to colour. Whatever the cause of chromatic aberration may be, it is not produced by reflection at any surface, plane, spherical or of any other form, as Newton soon discovered. Hence came Newton's decision to relinquish refracting for reflecting telescopes, especially as three years before then, in 1663, James Gregory of Aberdeen had already proposed this type of instrument. But Gregory was in complete ignorance of the cause of the imperfection of refracting telescopes, and in any case he never made a reflecting instrument at all. Newton discovered the cause—which had baffled the mightiest intellects of the generation before him—when he was a graduate of one year's standing, and made a reflecting telescope with his own hands about two years later.

The three subjects to which the youthful Newton devoted the years immediately following his graduation, and in which he made such remarkable discoveries so early in his career, were an epitome of his life's work. Fluxions, gravitation and optics—these were the main concern of the whole of his scientific life. For over thirty years he continued to achieve conquest after conquest. A decade he devoted largely to the prosecution of his optical researches and to the consolidation of the great discovery of the spectrum. A second decade he devoted to the working out and publication of his still greater discovery of the fact and the law of universal gravitation, the discovery that Newton is most often remembered for and in which his influence is still alive to-day. The subsequent decade of his scientific life at Cambridge Newton devoted to the further development of his astronomical work, and the consolidation of his pure mathematical researches, notably the method of fluxions. No other man in the history of science presents a record of devoted service in the cause of science with such conspicuous genius and success.

CHAPTER VI

THE OPTICAL DECADE, 1668-1678

> Hail, holy Light! offspring of Heaven first-born ...
> Bright effluence of bright essence increate!
> Or hear'st thou rather pure ethereal stream,
> Whose fountain who shall tell?
>
> JOHN MILTON: "Paradise Lost," Book III

NEWTON returned to Cambridge early in 1667. The latest recorded absence due to the Plague was Lady Day, 1667. He said nothing to anybody about his discoveries and speculations in mathematics, astronomy and optics. His merit was obvious, however, and on October 1, 1667, Newton was elected Minor Fellow of Trinity College. He could now settle down and work without any serious economic worries, and devote all his attention to the sciences.

There is reason to believe that as Fellow of Trinity Newton was assigned rooms in the staircase which has since become famous as Newton's staircase, east of Great Court. The rooms he occupied then were not those pointed out nowadays as Newton's rooms, but rather the ground floor apartments abutting on the College Chapel. The so-called Newton's rooms, on the right-hand side on the first floor, were occupied by him later on.

Newton was soon busy at his researches—with drills and hammers, magnets and compasses, prisms and materials for polishing glass and metal. He spent a few months in Lincolnshire, staying there in December over Christmas, 1667, and into the month of February, 1668: but he returned to Cambridge and took the

THE OPTICAL DECADE, 1668-1678

degree of M.A. in March. He was elected Major Fellow in the same year. He spent August and September in London, no doubt for the purchase of materials for his optical and other, especially chemical, experiments, and then returned to Cambridge in order to execute the design of more than two years' standing, namely, to make a reflecting telescope.

James Gregory had proposed a reflecting instrument consisting of two concave mirrors turned towards one another. Light from the object to be observed was to be reflected from one of the mirrors—the wider one —to a focus in front of the smaller mirror. The latter would reflect the light again and bring it to a focus through a hole in the first mirror so as to be observable by a person placing an eye-piece behind this hole. Newton very much disliked the hole in the larger mirror, and preferred instead of the second concave mirror a plane mirror placed at 45° with the axis of the telescope. The rays would thus be reflected to a focus through a hole in the side of the telescope tube. Newton's idea is the one usually followed nowadays, though a variant of Gregory's form, in which the small concave mirror is replaced by a small convex mirror— the hole in the large mirror and eye-piece behind it being retained—is represented by one very large reflector in existence to-day. This form is known as Cassegrain's telescope : Cassegrain was a contemporary of Gregory and of Newton.

There is not really very much to choose between the different forms, as Newton himself admitted. Newton's real merit consists in the fact that he actually made such an instrument—the first reflector in history—towards the end of 1668. It was not a great instrument either in size or in performance. Its aperture was one inch and its length was six inches. Newton used a magnification of forty times, and claimed that this was as much as could be done with a refracting telescope six feet long. He turned the telescope to the sky and saw with his tiny instrument the disc of Jupiter and the then known four satellites very nicely, and the phases of Venus with a little difficulty.

While Newton was working at this instrument, and engaging himself in researches of a chemical nature—he spent quite considerable sums of money on chemicals, furnaces, and a "theatrum chemicum" in April, 1669—the turning point in his career was close upon him. Isaac Barrow, a famous preacher and an excellent mathematician, had had Newton under observation for several years. Barrow was the first Lucasian professor of mathematics in the university, and had contributed with distinction to the development of geometry and analysis. In 1669 Barrow decided to devote himself entirely to theology and to resign the Lucasian chair. A successor was needed, and with remarkable vision Dr. Barrow nominated the twenty-six year old Newton for this post. No doubt Barrow was influenced to do this by two indications he had received of Newton's learning and ability. The first was Newton's help in Barrow's publication of his lectures on optics in 1669. The second was a paper that Newton gave to Barrow in the same year, entitled "On Analysis by Equations with an Infinite number of Terms," the first written statement on Newton's discoveries in mathematics, particularly fluxions, that we know to have found its way into another man's hands. Newton suggested that the paper might be sent to John Collins, a contemporary mathematician who corresponded with many mathematical workers and acted as a sort of liaison between them. Barrow received the paper in June, 1669, he sent it to Collins on July 31st, and three weeks later Barrow wrote again to Collins expressing pleasure at the favourable opinion that he had formed of the paper and divulging

THE OPTICAL DECADE, 1668–1678

that the author was Isaac Newton. Collins took a copy of this paper and returned the original. It is remarkable that Barrow and Collins, more experienced men than Newton, also failed to realize the importance of publishing the paper, both for Newton's own reputation as well as for the advancement of learning. It was not till forty-two years later that the contents of this manuscript were given to the world. In this way Newton's priority as the discoverer of the fluxionary method was allowed to come under a shadow of doubt, and the seeds of future unpleasantness were sown.

Newton's brilliance was recognized in Cambridge, and he was appointed Lucasian professor on October 29, 1669. As professor Newton had to lecture once a week one term each year on some mathematical topic, and to devote two hours per week to audiences with students who wished to consult the professor about their work and any difficulties they might have encountered. Newton chose to lecture on optics, for he was then in the midst of his further researches on the spectrum, the results of which he laid before his students. The lectures as such were not published till very long afterwards, when Newton had already left Cambridge and relinquished the professorship. Fortunately, however, his optical work was to come before a wider audience of scientific men in a somewhat peculiar manner. Later he lectured on algebra and gravitation.

Newton himself had no very high opinion of his little reflecting telescope. He made a second instrument, of about the same size as the first, but slightly better. The Royal Society, which had been founded at the instance of King Charles II in 1667, got to hear of the actual existence of a reflecting telescope, and sent a request to Newton that he might send them the instrument for inspection. It seems that Newton was a little vainer about his practical workmanship than his theoretical researches, for he readily acceded to this request. The instrument was sent to London, and not only the Royal Society, but also the King, saw it and expressed admiration for it. This telescope still exists as one of the most treasured relics in the possession of the Royal Society to-day.

The reputation of the telescope maker began to spread. He was at once proposed for election to the Royal Society, and elected without delay on January 11, 1672. This compliment pleased Newton very much, and he promptly offered to communicate to the Royal Society what he considered to be far more important and valuable, namely, his discovery of the spectrum and of the nature of colour. Newton's own opinion of his discovery is interesting in view of his general modesty and soberness of expression : he refers to it as " being in my judgment the oddest if not the most considerable detection which hath hitherto been made into the operations of nature " !

On February 6, 1672, Newton sent his paper to Oldenburg, the Secretary of the Royal Society, and it was published soon afterwards. By then Newton had reached more definite conclusions about the mysterious five-fold extension of the section of a beam of light on passing through a prism. At first Newton looked for various and quite impossible ways of explaining this result. But the conviction soon dawned upon him that there must be in the divergent coloured beam a number of beams of different colours, which had suffered different deviations by refraction through the prism. He found that the coloured patch could be made white again and the section restored to its original circular shape by taking the divergent coloured beam through another exactly equal prism, but with the bending taking place in an opposite sense to the refraction through the first prism. He then examined the light falling on various portions of the extended

THE OPTICAL DECADE, 1668-1678

patch on the screen. He was able to prove quite clearly and conclusively that these lights were all different from one another, and did actually suffer different refractions in going through the prism. Red, orange, yellow, green, blue, indigo, violet were the colours thus distinguished, the bending suffered through the prism being least for the red light, and increasing till the greatest bending was suffered by the violet light.

This was indeed a " considerable detection . . . into the operations of nature." White light is not homogeneous : it consists of differently coloured rays of varying refrangibility, or liability to be bent in passing from one medium into another. A convex lens will therefore not produce one white image, but a number of coloured images, a violet one nearest the lens, a red one furthest from the lens, and the other colours ranging in between. The cause of the imperfection of telescopes became quite obvious.

All this Newton set out in his paper to the Royal Society. He justified in this way his making the reflecting telescopes, and suggested that reflecting microscopes might also be made. Let it be said at once that the reflecting telescope of Newton was but the forerunner of mighty instruments based upon his construction. The hundred-inch reflector at Mount Wilson Observatory in California is a direct descendent of Newton's tiny telescope, through a long and noble lineage of reflecting telescopes made by William Herschel. Nothing considerable ever came, however, of the suggested reflecting microscope.

But Newton went further. Colour became to Newton a definite property of a ray of light, immutably associated with the amount of refraction that it undergoes in passing from medium to medium. He attributed the colour of a body to the light by which the body is seen and not to the body itself, and was thus the first man to give a reasonably accurate account of this important matter.

We have already seen that new ideas are not easily assimilated at any time, and this was particularly the case in the seventeenth century. Even men of science who were themselves original workers of great merit, failed to recognize in Newton's paper one of the finest contributions ever made to the theory of optics. Men of insignificant scientific stature raised objections to Newton's experimental results, and he answered carefully and with a good humour. This became irksome, and Newton began to lose patience. But Oldenburg induced him to go on, and to spend much time and energy in correcting his adversaries, and detecting their errors. Considerable discussion arose over the question of the five-time extension of the section of the beam. It was claimed by Lucas of Liège that he had carried out the same experiment as Newton and had obtained an extension of only three and a half times. Both Newton and his adversary maintained the accuracy of their measurements. There can be little doubt that both were right. The amount of " dispersion " depends upon the nature of the prism—the kind of glass used. Some glasses give a much larger dispersion than others, for the same amount of general bending of the beam. This was not suspected by Newton. Had Lucas been fortunate enough to induce Newton to go into the question of the kinds of glass used by the two experimenters then a discovery of capital importance would have resulted.

Newton's conviction that dispersion is always proportional to general deviation led him to conclude that it was quite impossible to eliminate chromatic aberration in a refracting instrument. This is now known to be untrue. By combining a convex lens of crown glass with a concave lens of flint glass we can make an achromatic combination, i.e. a combined lens

THE OPTICAL DECADE, 1668-1678

which gives an image almost (but not quite) devoid of colour aberration. If the mighty reflectors of to-day are monuments of Newton's genius in optical discovery, the mighty refractors of to-day are equally imposing monuments of the imperfection of Newton's knowledge and the harm that accrues to science when the authority of a fallible human being is allowed to silence criticism and to discourage further research. It was only several generations after Newton's insistence upon the amount of dispersion being so-and-so, without paying any attention to the material of the prism, that his mistake was discovered and rectified. Since then refracting telescopes have become the really accurate instruments of astronomy, the mainstay of astronomical observation.

In his account of the spectrum and the nature of colour Newton did not make any assumption as to the nature of light itself. He was content to speak of it as "something or other propagated every way in straight lines from luminous bodies." The experimental facts that he had discovered seemed to him to transcend any theory that one might invent concerning the ultimate nature of light propagation. When Robert Hooke, an Englishman of great ability whom we have already mentioned, and Christian Huygens in Holland, one of the most famous scientists of that century, raised objections to the spectrum on the ground of the wave-theory of light which they believed in, Newton felt that they were introducing irrelevances, and from men of their scientific power this was annoying. He was always somewhat unwilling to face publicity and criticism, and had on more than one occasion declined to have his name associated with published accounts of some of his work. He did not value public esteem as desirable in itself, and feared that publicity would lead to his being harassed by personal relationships—whereas he wished to be free of such entanglements. The effects of his publishing his optical results therefore decided Newton to desist from such publication in the future. "I see I have made myself a slave to philosophy. . . . I will resolutely bid adieu to it eternally, excepting what I do for my private satisfaction, or leave to come out after me; for I see a man must either resolve to put out nothing new, or to become a slave to defend it." Fortunately, Newton did not persist in this determination—except for the fluxionary method and other pure mathematical work, which seem to have been doomed to remain the private possession of Newton and a few of his friends for a long time to come.

We know very little of Newton's personal life throughout this period. We have, however, information on the authority of a close friend who shared rooms with him at Trinity, that Newton developed very early in life that forgetfulness about food and a carelessness about his own health which became characteristic of him in later life. He would often work till very late and miss restful sleep because of his work. At the age of thirty he was already turning grey. He suspected himself to be inclined to consumption and made medicines for his own use. Newton was indeed fond of dosing and curing himself. "His breakfast was orange peel boiled in water, which he drank as tea, sweetened with sugar, and with bread and butter." When Newton caught a cold he took to his bed and cured it by perspiration. All this reads like the life of a bachelor, far from the help of mother or wife, doing things for himself in a clumsy way and living for the work on which he is engaged.

Newton was a kind man, sincerely pious and an influence for good in many ways. As regards his financial position, though he obtained various sums of money from his mother before being elected to the Lucasian chair, we may surmise that the scanty

THE OPTICAL DECADE, 1668-1678

yield of the farms at Woolsthorpe and Sewstern was not overmuch for the needs of Mrs. Smith and the three children by her second marriage. Newton was thus left to his own resources, which were not very great as his chair was only moderately endowed. He was also very free with his money, bestowing loans and gifts upon relations and friends with considerable liberality, and spending freely upon his scientific apparatus.

It therefore became a matter of considerable concern to Newton that his College Fellowship was soon to expire, so that he would have to live upon the stipend of the Lucasian chair alone. In order to continue to hold his fellowship Newton would have had to go into holy orders. He declined to do this, and instead made an application to the King for a dispensation allowing him to retain the fellowship as a layman and as Lucasian professor. The petition was granted in April, 1675, an act of royal grace which does much to whiten the character of the merry monarch.

Why did Newton refuse to go into orders? He was a man of excellent character, sincerely and profoundly moral in his opinions and in his practices. He believed in the fundamentals of religion and even, as we shall see, wrote much on theological matters. It is clear that Newton desired as complete freedom in theological thought as in scientific belief, and refused to be bound in any manner. He did not accept without questioning all that was taught by the prevalent churches. His faith was that of a profound and independent thinker and investigator, seeking for the causes of all effects and for the unifying cause underlying all things, rather than that of an implicit believer in doctrines laid down by ecclesiastical authority.

These economic disturbances of Newton's equanimity are mirrored in a remarkable manner by his relationship to the Royal Society. A fellow of this body had to pay an entrance fee of two pounds, and a continuous subscription of one shilling per week. In March, 1673, Newton wrote to the secretary Oldenburg resigning his membership. He gave as his motive the fact that he lived too far from London to be able to take full advantage of the meetings of the society. Oldenburg must have known better, for he offered to get Newton excused the shilling per week payment, and Newton himself raised no objection to an application to this effect being made to the Council of the Royal Society. Oldenburg did not present the application at once; possibly he waited to get a few more such applications to be dealt with at the same time. In January, 1765, he mentioned "that Mr. Newton was now in such circumstances that he desired to be excused from the weekly payments," and Newton and others—including Robert Hooke—were granted this relief. It is a pleasing thought that the greatest minds of the day had to submit themselves to what must have been felt as a humiliation in order to be relieved of the payment of one shilling per week!

The scientist does not work for reward. The impelling motive is not personal gain or even personal fame: it is the insistent desire of the healthy human mind to know how Nature manifests herself around us, and the causes of these manifestations. Scientific men and women are continually giving of their best in the onward march of knowledge; they place at the service of the community their powers and their energies. By membership of learned societies they even pay for the publication of the results which they deduce, often at their own expense and without interest or encouragement from anybody. Surely a community that has so much to spend on personal luxuries, and on fashionable stars in the theatre and on the screen, could devote a little more to the fostering of scientific research and to making the task of the worker easier by removing all anxiety of economic embarrassment? Two and a half centuries have passed since then and some improvement has taken place in this respect, but Great Britain has much leeway to make up even now.

The discovery of the spectrum was one of the most important events in the history of optics, and indeed, in the history of science. It has led to developments in all branches of natural knowledge. The discovery of the black lines in the solar spectrum early in the nineteenth century gave us spectrum analysis, which has revolutionized physics, chemistry and astronomy. Modern astronomy, rightly called Astrophysics, consists essentially of the application of the spectrum to heavenly phenomena. By its means we discover the constitution of the stars, their motions towards us and away from us, their distances when too remote to be measured directly, the events on the sun's surface which reveal electro-magnetic fields and disturbances of tremendous significance. The spectrum has by the work of Bohr and others given us an insight into the ultimate constitution of matter, and led to the initiation of the quantum theory, the implications of which we can as yet hardly guess. The spectrum has helped to establish Einstein's gravitational theory; and led to the discovery in the heavens of matter as immensely denser than the matter on our earth as the giant stars are lighter than our earthly matter.

It is not possible in this book to give a detailed account of all Newton's optical work, and a mere catalogue of discoveries will convey little to the general reader. Yet a few words on some of Newton's work will illustrate the magnitude of his powers and the influence he exerted on subsequent developments.

Newton not only applied the reflecting principle to the microscope, a suggestion that has not been very fruitful, but he also made the capital proposal that it would be an advantage when using ordinary refracting microscopes to employ light of one colour, or monochromatic light. While working with the plane mirror in his reflecting telescope Newton saw how imperfect reflection from a metallic surface is bound to be, and he hit upon an idea without which the powerful modern binoculars would be impossible. This was that in view of the high refractive index of glass, an isosceles right-angled prism of glass could be used as a reflector, giving much better results than metals.

Discoveries of great theoretical importance were communicated to the Royal Society at the end of 1675 in connexion with the colours of thin films of transparent matter. It is of course well known that very thin plates of transparent matter give different colours. The beauty of the soap-bubble has been immortalized in a famous picture. What are these colours due to? Robert Boyle, famous for his law connecting the volume and pressure of a gas, and Robert Hooke, observed the phenomena in mica plates, but could not discover the relation between the colour produced and the thickness of the plate because of the extreme minuteness of the latter. Newton hit upon a beautiful and ingenious idea which we still use in our physical laboratories in Newton's rings and in other applications.

The tangent to a circle at any point hugs the circle, and the deviation of the circle from it is very small even for considerable distances along the tangent, if the radius of the circle is large enough. If A is the point of contact of the tangent and P is a point not too far away on the circle, while PT is perpendicular to the tangent AT at A, then $TP = TA^2/2a$ where a is the radius, supposed large. Thus, if the radius is 50 feet and TA is one-twentieth of an inch, we get for TP 1/480,000 of an inch. If TA is one-tenth of an inch, TP is 1/120,000 of an inch. If TA is an inch,

TP is 1/1200 of an inch. If then we take a flat glass plate and put into contact with it the spherical surface of a lens, with radius 50 feet, we have within a distance of one inch from the point of contact a film of air which varies from zero at the cenree to 1/1200 of an inch at the edge. Not only is the film thin enough to produce

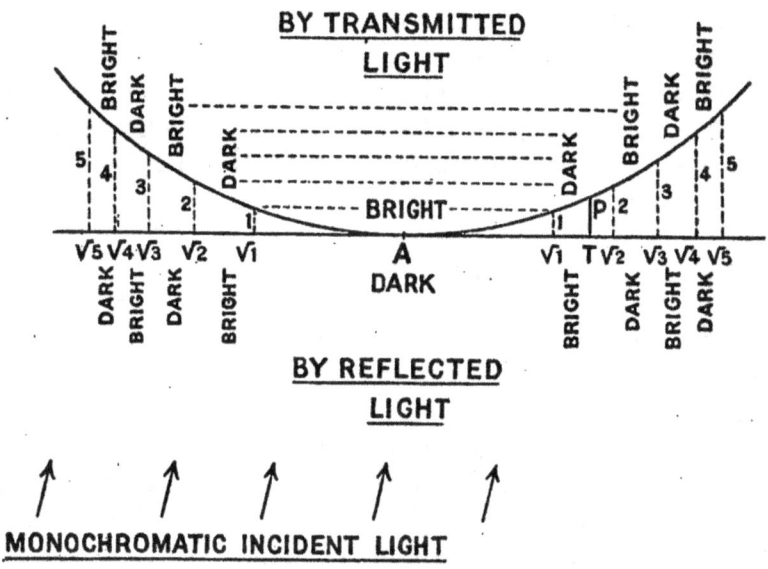

FIG. 6
NEWTON'S RINGS
Alternate bright and dark rings by transmitted light; alternate dark and bright rings by reflected light; produced by thin air layer between a flat and a curved surface.

colours like those of the soap-bubble, but we can also calculate the thickness of the film at any measured distance from the point of contact.

When Newton observed white light transmitted through the film or reflected from the film he saw coloured concentric rings, in which the colours were by no means so simply arranged as in the spectrum.

On using light of one colour, however, the appearance became very much simplified, less beautiful of course, yet easier to understand. Newton found that by reflected light of any particular colour he got a circular dark patch in the centre, then a ring of that colour, then a dark ring, then another ring of that colour, then a dark ring again, and so on up to about twenty rings. With red light the successive radii were larger, with violet light they were smaller. Newton also found that the transmitted light gave a circular light patch in the middle, then a dark ring, then a light ring, then a dark ring, and so on, the radii being exactly the same for transmission and reflection in the case of any given coloured light, but lightness and darkness interchanging from one to the other. The radii were found to be proportional to the square roots of the successive numbers 1, 2, 3, 4, etc., so that the air thicknesses at the successive circumferences were proportional to the numbers 1, 2, 3, 4, etc. themselves.

Newton at once concluded that a ray of light cannot have exactly the same properties at all points along its length. There must be variations in these properties, "fits" as Newton called them. It is clear that a ray of light of any given colour prefers to be transmitted for a certain—very short—distance depending on the colour itself, then prefers to be reflected for an equal distance, then prefers to be transmitted for such a distance, and so on indefinitely. A ray of light thus has a space-period depending upon its colour, shortest for violet, getting longer for indigo, blue, green, yellow, orange, and being longest for red.

One can think of many objections to this theory of fits; but the most disconcerting feature about this matter is the way in which this experimenter of genius failed to grasp the (now) obvious implication of his beautiful experiment. Surely there must be a periodi-

THE OPTICAL DECADE, 1668–1678

city along a ray of light propagation, and this must be in the form of a wave, colour being in fact a property associated with the space-period or wave length? Why did not Newton postulate a wave-theory and use the great principle of interference? No great originality was required, since both Hooke and Huygens had used the wave-theory in their controversies with Newton, and Hooke had even announced the principle of interference.

Strange as it may appear to us now, Newton refused to accept a wave-theory of light. Did the opposition of Hooke and Huygens to Newton's discovery of the spectrum, based upon their wave-theory, prejudice Newton against such a view? It is hardly credible, for Newton was too honest a thinker to be affected in this way. The fact is that Newton was severely practical in his conceptions. He always desired to see or feel, so to say, any natural force or other agency before he would accept it as an explanation of observed phenomena. His was the kind of mind that would interest itself in proving by direct observation the daily rotation of the earth and its annual revolution, that would have welcomed with tremendous joy Bradley's discovery of aberration as proving the latter and Foucault's pendulum as proving the former.

Of course Newton was curious to explain why these fits existed and he did in fact invoke an " ether " to explain this and other phenomena. He supposed that a fluid ether permeates all space, whether occupied by matter or not, but that the density of the fluid varies from point to point. In this way Newton explained many properties of matter. As regards light he took it to be due to the very fast propagation of very tiny corpuscles, which are acted upon by the particles of the ether and so receive this property of alternate fits of preference for transmission and for reflection. This view he expressed in 1675 and he developed it a few years later in a letter to Robert Boyle. Apparently, however, an ether the existence of which he could not prove directly, was not palatable to Newton, and he rejected the ether—to be revived later in query form when universal gravitation was added to the things requiring explanation.

By means of his theory of fits Newton was able to make considerable advance in the explaining of transparency and opacity, and of the colours of different bodies. Transparency and opacity depended, according to Newton, upon the sizes of the spaces between the ultimate particles of a body, while the colour of the body depended upon the sizes of the ultimate particles themselves. It would take us too far to discuss these views in detail. One thing we can say, however: they represented the first rational set of opinions ever put forward on the subject, and were a tremendous advance on the views held at that time by the greatest of physicists and mathematicians.

Another important optical experiment carried out by Newton referred to what we nowadays call diffraction. Light is supposed to travel in straight lines so long as no refraction takes place. Hence, in uniform and still air no deviation from rectilinear propagation should occur. This is not the case. Light rays do bend round a corner just as sound bends round a corner. In the case of light, however, the bending can be detected only by means of very careful observation. An Italian, Grimaldi, first obtained the bending and the result was published in 1665. In 1674 Hooke carried out further experiments and betrayed his genius by suggesting the principle of interference in order to explain the phenomenon. Newton carried out diffraction experiments at about the same time. He cast the shadow of a human hair upon a screen and found

THE OPTICAL DECADE, 1668-1678

the shadow to be much wider than the thickness of the hair, and to exhibit coloured fringes. He did not accept Hooke's explanation—which is the right one—and attributed the phenomenon to the possibility of flexibility in a ray of light. Rays of different colours have different flexibilities. As a ray passes the edge of a body a wave-like effect is produced on the ray, giving it a shape like that of an eel! Rectilinear propagation is upset, and colours are introduced. This queer explanation was published nearly thirty years later in Newton's "Optics."

Why Newton should have been reluctant to accept the wave-theory of light, while prepared to postulate these eel-like motions, it is difficult to explain. Personal idiosyncrasy is as much the prerogative of genius as of lesser minds. Newton filled the universe with an ether in order to account for the fits in a remote and indirect manner. Yet he declined to follow Huygens and Hooke in the wave-theory and in interference.

The consequences to science were very unfortunate. When Newton had become famous as the discoverer of the law of universal gravitation and had explained the motions of the planets in a manner unique in scientific history, the prestige of Hooke and of the great Huygens could not prevail against his renown. The wave-theory of light fell into disrepute. It was not till more than a century later that the theory was rehabilitated by the work of the Englishman Young and the Frenchman Fresnel. The hero-worship accorded to genius has its vicious as well as its virtuous side. For genius is but human after all: it is liable to honest mistake and even to prejudice and to passion. Science should stand above all human considerations. It must not bend the knee to any human authority. It must rest on honest research, based upon complete intellectual independence. The raw schoolboy may be right where the sage of mighty repute is mistaken.

The weight of Newton's authority delayed optical progress for many generations. It was not Newton's fault, of course; it was the fault of the lesser men who followed him and mistook idolatry for sincere admiration.

CHAPTER VII

THE GRAVITATIONAL DECADE, 1678-1687

> If with giddier girls I play
> Croquet through the summer day
> On the turf,
> Then at night ('tis no great boon)
> Let me study how the moon
> Sways the surf.
>
> MORTIMER COLLINS : " Chloe, M.A."

WE now enter upon the second or gravitational decade of Newton's scientific career. The period between 1668 and 1678 was not exclusively devoted to optics. His famous paper on the method of infinite series and fluxions was written in 1669. He wrote a second treatise on fluxions in 1672, but this also remained unpublished. He communicated with Huygens concerning gravitation in January, 1673, and he wrote a paper for the Royal Society on the excitation of electricity in glass in 1675. In the same year Newton wrote to Nicholas Mercator, a German Mathematician then resident in London, concerning the librations of the moon, the fact, namely, that although on the whole the moon always turns the same half of its surface to the earth, we do notice slight changes in this respect, as if the moon were oscillating from one side to the other. Newton explained this satisfactorily as being due to the uniform rotation of the moon about its axis in exactly the time it takes to complete a revolution round the earth—this motion of revolution is not uniform, so that sometimes the moon has apparently rotated too much, at other times too little, in relation to the angle through which it has revolved round the earth.

In the midst of his scientific preoccupations Newton even found time to discuss agricultural matters. He was perhaps making inquiries on behalf of his college, but in September, 1676, he wrote to Oldenburg on the question of planting apple trees for the manufacture of cider! He wished to know: "What sort of fruit are best to be used, and in what proportion they are to be mixed, and what degree of ripeness they ought to have? Whether it be material to press them as soon as gathered, or to pare them? Whether there be any circumstances to be observed in pressing them? or what is the best way to do it?"

In the same year Newton wrote to the great German philosopher and mathematician, Leibniz,, concerning his method of fluxions. Leibniz sent a reply in which he gave an account of the differential calculus with the notation in vogue to-day, and which Newton must at once have seen to be the same thing as his own fluxions, differing from his invention only in the notation.

But Newton's attention was soon directed to astronomy and gravitation. We may assume that the abortive calculations made at Woolsthorpe in 1666 never quite dropped out of his memory. The problem of planetary motions was too insistent and the failure too tantalizing to leave a man like Newton quite unaffected. Meanwhile another man had been writing on the subject, Robert Hooke.

Hooke was a man of great genius, but owing to his preoccupation with all sorts of matters he never quite finished off any great line of research, and consequently never reaped the full fruits of the seeds that he sowed. Newton was on the other hand a remarkably successful investigator of singular stability of character and steadiness of application. We have already seen that Hooke and Newton had had occasion to dispute over

THE GRAVITATIONAL DECADE, 1678-1687 83

optical matters. Hooke had also treated the younger Newton somewhat unkindly on the occasion of his first communication to the Royal Society, when Newton's reflecting telescope was under inspection. Hooke may have seen in Newton the younger and greater rival who did the things that Hooke might have done if not anticipated by Newton. A mild jealousy of Newton grew up in Hooke's mind, and perhaps a slight suspicion of Hooke entered Newton's mind.

Similar unfortunate circumstances arose soon in connexion with gravitation. While Newton was speculating on his failure at Woolsthorpe, Hooke sent to the Royal Society, in 1666, a paper on the question of how the earth's gravity varies with height above the earth, making the important suggestion that the force of gravity could be measured by means of a " swing clock " or pendulum. He followed this up with another paper in the same year, suggesting that the planetary motions could be explained by means of an attraction towards the central body. This vague idea, which had occurred to several others before then, took more definite shape in a third publication which appeared in 1674. In this work Hooke quite definitely postulated universal gravitation, stating that each heavenly body not only keeps itself together by its own attraction, but also exerts a force upon every other heavenly body. He did not yet know what law to use, how the gravitational force exerted by a body depended upon the distance from this body; but in 1679 Hooke wrote to Newton and made the assertion that if the earth's gravitation varied inversely as the square of the distance the curve described by a projectile would be an ellipse, with a focus at the centre of the earth.

In spite of all this, however, Hooke's claim to consideration in the question of priority in this matter could have been of little weight. Newton had discovered the inverse square law thirteen years before by direct deduction from Kepler's third law of planetary motions. He had been held up by a quantitative discrepancy which turned out to be due to a quite irrelevant circumstance, namely, his ignorance of the correct size of the earth. Hooke's suggestion that the inverse square law would yield elliptic motion round a focus was not accompanied by a proof. It was Newton who established independently, and by brilliant mathematical reasoning, that all the facts of planetary motion then known could be accounted for by universal gravitation according to the inverse square law. By universal consent Newton was acclaimed as the discoverer of the secret of heavenly mechanics. This reacted upon Hooke's mind with some unpleasant consequences.

If Newton had been acquainted with correct values of the earth's radius published thirty years before his Woolsthorpe calculations, would he have made his great gravitational achievements much earlier? Possibly in his Lincolnshire exile Newton had no means of referring to books on the subject, but when he did get back to Cambridge he should have become acquainted with the correct radius of the earth. Accurate measurements by the Frenchman Picard were communicated to the Royal Society on January 11, 1672, the very meeting at which Newton was elected a fellow and at which his reflecting telescope was exhibited, and published in the Transactions in 1675 : Newton cannot have missed seeing these results. It is a remarkable fact nevertheless that he did not take up the matter again until his attention was directed to it by Hooke himself.

Robert Hooke became Secretary of the Royal Society in 1678, after Oldenburg's death. He wrote to Newton asking him if he had something new to communicate to the society. Newton replied with a dissertation on

THE GRAVITATIONAL DECADE, 1678–1687   85

the diurnal rotation of the earth. The reason why the earth's rotation remained undiscovered till so recently, and why some cranks still doubt it to-day, is simply

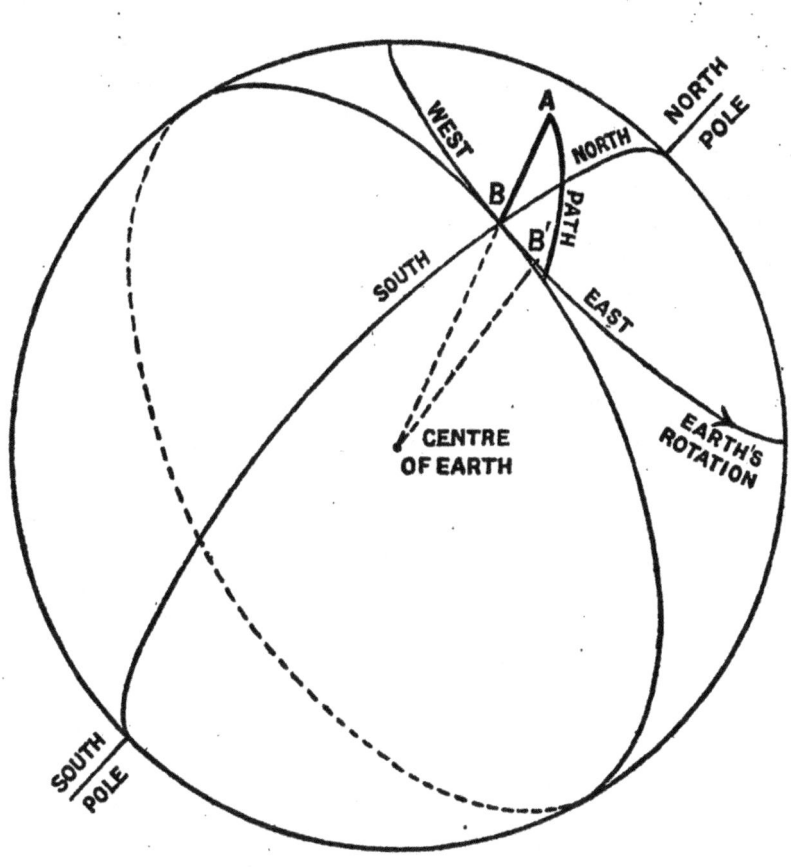

FIG. 7
FALLING BODY AND ROTATING EARTH

Path of stone dropped from A lies in that plane through the centre of the earth which touches B's circle of latitude at the point B. The stone falls east of B' (the new position of B when the stone reaches the ground) and south of it.

that this rotation takes place in a perfectly smooth manner, with all the things on the earth accompanying it in its rotation. No differential motions are visible to us : the solid earth rotates, the waters of the ocean, the air, houses, trees, we ourselves—all goes with the earth. How can the supposed rotation be distinguished from a state of rest of the earth ?

It was Newton who saw that this very fact that all things on and near the earth accompany it in the diurnal rotation could be used to demonstrate the rotation ocularly. For consider a point A vertically above any point B of the earth's surface, so that AB is along the extension of a radius from the earth's centre. In the diurnal rotation the line AB is carried with it bodily ; hence the point A has a velocity, due to this rotation, greater than that of B, since it is further than B from the common axis of rotation. Now let a body be dropped at A. During the fall it receives downward acceleration taking it to B. But it also had initially a horizontal velocity west to east greater than that of B. Hence the falling body will not reach the earth exactly on B, but slightly to the east of B. The deviation would be very small indeed : for AB = 100 feet, the deviation in London would be $\frac{1}{77}$ inch. Newton suggested this as a possible experiment—a straightforward test of the earth's rotation.

The suggestion was adopted and Hooke was appointed to carry it out. Before doing so, however, he examined the matter more carefully and came to the conclusion that Newton was not quite right. If B is on the Equator, then Newton's argument is faultless. But if B is not on the Equator, then the deviation will not be exactly to the east : it will be also a little to the south in the northern hemisphere, to the north in the southern hemisphere. It is not difficult to understand why this should be the case. The body during its fall is pulled along AB. Relative to B the initial motion of A is to the east. Hence the body's fall takes place in the plane containing the line AB and touching the cone described by AB, along the line AB. This tangential plane lies outside the cone and so cuts the earth in a circle which lies outside the parallel of latitude at B : in other words, the body must come a little south (or a little north) as well east, parallel of latitude that defines east and west.

Newton had missed this point, and at once admitted it when his attention was drawn to it. Hooke carried out the experiment and verified the deviation in December, 1679, the first ocular proof in history of the rotation of the earth ! It was the year of the Habeas Corpus Act.

In connexion with this investigation the question arose between Newton and Hooke as to what the path of a body actually is under the earth's attraction, taking this attraction to vary inversely as the square of the distance. Newton conjectured that the path would be a spiral curve. Hooke dissented from this and asserted that the path would be an ellipse with a focus at the centre of the earth. Newton himself said about this : " And tho' his (Hooke's) correcting my spiral occasioned my finding the theorem, by which I afterwards examined the ellipsis : yet am I not beholden to him for any light into the business, but only for the diversion he gave me from my other studies to think on these things, and for his dogmaticalness in writing, as if he had found the motion in the ellipsis, which inclined me to try it, after I saw by what method it was to be done."

It is clear that Hooke could not prove his assertion : it was a guess, based upon his extending the idea of Kepler's laws of planetary motion to the motion of a projectile under the earth's gravitation. For this Hooke deserves great credit, yet the real discoverer was Newton, for it was Newton who turned Hooke's useless guess into a certainty, being impelled thereto by Hooke's " dogmaticalness in writing." Newton was able to construct a *proof* of the proposition that a body moving under an attraction towards a fixed point varying as the inverse square of the distance from this point, describes an ellipse with a focus at the centre of attraction.

This was a result the importance of which could hardly be exaggerated, and Newton should have proceeded now to complete a theory of the solar system. But, having proved the elliptic motion, Newton " threw the calculations by, being upon other studies ; and so it rested for about five years." He was making alloys for metallic mirrors—one can only wonder whether they were important enough to make Newton " throw by " the calculations upon which his greatest fame in human history was destined to rest. He corresponded with the Astronomer Royal, Flamsteed, about the great comet of 1680, and the views Newton then expressed make it clear that he had not yet reached a correct understanding of cometary motions. If he had paid more attention to the train of thought roused by Hooke, he would not have held these erroneous views.

No apology need be offered for Newton's apparent indifference to the matter. Evidently he was still under the influence of his Woolsthorpe failure to explain the moon's motion by the earth's gravity. He could not deal with the motions of the planets under an attraction towards the sun, without dealing at the same time and on the same principle with the moon's motion round the earth. To Newton's practical mind the latter was even the more important aspect, for only in the case of the earth's gravity could he examine a really obvious and measurable force—a force that enters into all the phenomena of daily life. So long as the moon did not respond to his calculations he had no right to postulate universal gravitation as the cause of the planetary and lunar motions.

For some reason or another Newton's attention had

not been drawn to the more correct value of the earth's radius. He had taken it to be given by sixty English miles for a degree of latitude. The correct value is much more. Picard's value was 69·1 miles, a difference of over fifteen per cent. It would not have escaped the notice of a mind like Newton's that this was just the quantity required to correct his calculations of 1666. Yet it was not till June, 1682, that Newton became fully aware of Picard's measurements—or was it that only then did it strike him how these measurements would affect his calculations? We are told that at a meeting of the Royal Society in June, 1682, conversation turned upon Picard's work. Newton noted the result and on his return to Cambridge took up the Woolsthorpe calculations once again. Becoming conscious of the approaching almost perfect agreement between his theory and fact, Newton, the story relates, became so agitated that he was forced to entrust the arithmetic to a friend. Eventually the truth emerged. The force which drags the apple to the earth is the force that directs the moon round the earth. The daring suggestion that the heavenly bodies move under the same kind of forces and obey the same laws of motion as the earthy matter around us was vindicated. The road became quite clear for what may be described as the grandest generalization ever made in human thought.

Apparently Newton hardly ever published a discovery without being urged to do so by others. Even when he had arrived at the solution of the greatest problem that astronomy has ever had to face he said nothing about it to anybody. Fortunately other men were interested in the same problem. These were the three Englishmen, Robert Hooke, Sir Christopher Wren, excellent mathematician and brilliant architect, and Edmund Halley. Edmund Halley was twelve years younger than Newton, and a man of great ability, who, in 1684, had already done a number of important pieces of scientific and astronomical work, especially on comets. He became interested in the problem of gravitation and in January, 1684, he deduced from Kepler's third law that the sun's gravitation must vary inversely as the square of the distance—in other words, he did in 1684 what Newton had done in 1666. He tried to deal with the much more difficult problem of finding the path of a body moving under such an attraction, but he failed. He therefore sought out Wren and Hooke and asked them if they could supply what he could not achieve. Apparently both of them had deduced the inverse square law from Kepler's third law. But as regards the question of finding the path under such an attraction, Wren admitted he could do nothing with this bigger problem, while Hooke made the statement that upon the inverse square law " all the laws of the celestial motions were to be demonstrated, and that he himself had done it." Wren thereupon made a sporting offer that if Hooke or Halley could produce the mathematical proof, instead of the mere assertion, within two months, he would present the discoverer with a book of forty shillings. Hooke again said " he had it, but that he would conceal it for some time, that others trying and failing might know how to value it when he should make it public." But he never produced any such proof, and Halley, tired of waiting for it, decided to apply to Newton.

The momentous interview took place in August, 1684, about half a year before the death of King Charles II. Halley visited Newton at Cambridge, and came straight to the object of his visit. " What would be the path of a planet under a gravitational attraction varying inversely as the square of the distance ? " he asked. " An ellipse," came Newton's reply without hesitation. " How do you know ? " asked Halley.

"I have calculated it," answered Newton. Halley naturally wanted to see the proof, but Newton had apparently not paid much attention to the matter for some time and so could not immediately find his notes on the subject. He promised to let Halley have the proof soon.

Was Halley confident that Newton's proof would really be forthcoming? After all, his success with Newton was not very dissimilar from his success with Hooke, who had also claimed to have a proof and had not yet produced it. The fact that Halley took the trouble to go to Cambridge to see Newton shows, however, that already he had learned to admire Newton's genius and that he had complete confidence in Newton's ability to establish any claim he might make. In this he was not disappointed. Newton could not find his original notes, but he constructed the proof again—after a little trouble—and sent it to Halley within three months of the latter's visit to him.

Halley's joy was unaffected. He was a man who could see gladly the success of another greater than himself. He decided that such a discovery could not remain unpublished. He visited Newton again and found that the latter had meanwhile prepared a brief account of the motion of a body under a force directed to a fixed centre, particularly when the force varies inversely as the square of the distance, in which case the path is a conic with a focus at the centre of force. Newton had already lectured on the subject as Lucasian professor. Halley made Newton promise to send this short treatise "De Motu" to the Royal Society, a promise that was communicated to and welcomed by the society on December 10, 1684.

Great events were taking place in England then. Surrounded by intriguing parties who demanded opposite things, the King was in perplexity both as to his foreign relationships and as to the question of summoning Parliament to learn the will of his people. In the midst of it all, on February 6, 1685, the King died of apoplexy. He was buried in Westminster Abbey eight days later, and a perplexed people saw with mixed feelings the accession of Charles' brother, King James II. It has been said that the reign thus ended was the worst reign in English history: yet let us remember that Charles had founded both the Royal Society and the Royal Observatory, Greenwich, and that he had had the grace to do an act of decency to Newton in allowing him to retain his Trinity fellowship without going into orders.

It was in the midst of this national commotion that Newton wrote out his paper on planetary motions, and it was received by the Royal Society while the dead King was awaiting burial. Newton himself indicated that he was preparing a larger treatise on the subject, and in April, 1685, he began the preparation of his great work.

Newton was forty-two years of age and at the very height of his mental powers. If he showed greatness as a physical experimenter in his optical work, he was soon to prove himself even greater as a physical and mathematical thinker. Let us state clearly what was involved in the task he had undertaken. In the first place, the principles of dynamics had to be thoroughly grasped and clearly formulated, namely, that the absence of force or the balancing of the forces acting on a body means uniform speed in constant direction, without any acceleration, while any change in speed or in direction of motion of a body must be accounted for by a force or resultant of forces, proportional to the acceleration and acting in the same direction. Secondly, the type of force appropriate to the problem of planetary motion had to be considered and the inverse square law deduced. Thirdly, the physical reality of this attraction had to be proved, as presented

THE GRAVITATIONAL DECADE, 1678-1687

by the motion of the moon round the earth, the gravitational effects of which we experience daily and can measure accurately. Fourthly, the law of gravitation had to be applied in its general form to the planets and the motions as given by Kepler's laws accounted for accurately. For this purpose new and more powerful mathematical methods were required, and Newton had to invent them. Finally, the whole solar system had to be considered under the aspect of universal gravitation, and motions of satellites and comets explained, precession and tides brought within the ambit of scientific research.

A mighty task this was which could be accomplished only by a man specially favoured by the gods. Popular imagination has been captured by the third point, the association of the moon's motion with the fall of the apple—and rightly so, for this physical extension from the earthly to the celestial is one of the most significant steps in the revolution of ideas that Newton's work has brought about. But what exhibited Newton's powers to the full were the fourth and fifth points on which he laboured during the next two years.

Newton was not content with the astronomical observations of the older astronomers, and fortunately he found at hand a man prepared to give him the latest and best results. This was John Flamsteed, whom we have already mentioned as having corresponded with Newton a few years before about comets. He was a remarkable observer of the heavens and was appointed Astronomer to King Charles II at the early age of twenty-nine, in 1675. The Greenwich Observatory was built and Flamsteed headed the line of eminent astronomers that have adorned the post of Astronomer Royal. It was Flamsteed who gave Newton the best information possible about the orbit of Saturn, about the motions of the satellites of Jupiter and Saturn—the only satellites then known besides the moon—about the tides, about the flattened appearance of Jupiter's disc. For many years Newton depended upon Flamsteed for the facts upon which he built up the law of universal gravitation, and verified the consequences that flowed from the law.

The first part of the work was soon ready; it was the treatise " De Motu " more fully developed, forming Book I of the complete work. The second part or Book II was ready by the middle of the year 1685. Both books dealt with the theoretical aspects of dynamics, especially as required for application to astronomy. The third part, or Book III, was a much more difficult task, as in it Newton had to give detailed applications of his theory to the observed planetary motions and other phenomena in the heavens. Finally, on April 21, 1686, Halley announced to the Royal Society that " Mr. Isaac Newton has an incomparable Treatise of Motion almost ready for the press." On April 28th a manuscript entitled " Philosophiæ Naturalis Principia Mathematica " (The Mathematical Principles of Natural Philosophy) was presented to the society. As a matter of fact, this manuscript only contained Book I, but the importance of the whole work was speedily recognized and Halley was asked to report to the Council on the question of having the book printed. As the Council dallied in the question, the society resolved, at its meeting of May 19th, "That Mr. Newton's work should be printed forthwith in quarto." Still nothing was done, and on June 2nd the society adopted a most remarkable resolution: "That Mr. Halley undertake the business of looking after it, and printing it at his own charge, which he engaged to do " !

This is an interesting sidelight on the way scientific men who are enthusiastic about their subjects are prepared to face personal inconvenience and loss in the furthering of their aims. Halley was not a

THE GRAVITATIONAL DECADE, 1678-1687

particularly rich man and he had a wife and family. But when he realized the state of the Royal Society's finances he guessed that Newton's book would never be printed if it had to depend upon the society defraying the cost. Apparently the Royal Society had depleted its exchequer by printing a work on fishes, and in spite of the members' admiration for Newton's work they could not muster enough enthusiasm to give this admiration a practical form. Halley therefore decided that cost him what it may the book must be printed, and just as Halley had discovered Newton's genius in the matter of his gravitation theory, so it was Halley who took the risk of shouldering the financial burden involved in giving it to the world.

Halley was a distinguished man of science in his own right, and was to occupy the highest astronomical post in England, that of Astronomer Royal, and to give to the world the first prediction of the return of a comet. Yet, if Halley had done nothing more than secure the publication of the " Principia " he would have been assured of the everlasting gratitude of posterity.

If Halley in this way shielded Newton from financial worries, he could not shield him from other unpleasantness. Amidst the eulogies that greeted the presentation of Newton's manuscript, a discordant note had been struck by Hooke. We can easily imagine the feelings of Hooke at seeing the prize for which he so ardently yearned snatched from his grasp. While he had been speculating, Newton had worked and reached a happy conclusion of his labours. Hooke could not deny the great value of Newton's mathematical proof that the inverse square law of universal gravitation explains all the planetary motions then known, but he claimed that the idea to use the inverse square law had been given to Newton by himself.

This was really a foolish claim to make, since the

fundamental idea had occurred to several people before Hooke, and even in Hooke's time there were Halley and Wren who had deduced the inverse square law from Kepler's third law independently of himself and of one another. It was therefore very likely that Newton, whose genius was known to excel that of any one of this trio, had discovered this for himself, before Hooke had suggested the inverse square law to Newton in 1679. As we know, Newton made this discovery as early as 1666, and when told of Hooke's claim, Newton lost his equanimity and usual sense of fairness to such an extent as to retort that, so far from Hooke having suggested the inverse square law to Newton, Hooke had got the first idea of this law from others, and even from a letter that Newton had written to Huygens on the subject in 1673, through the secretary of the Royal Society, Oldenburg, and a copy of which Hooke must have seen when he succeeded Oldenburg. But Halley did all he could to prevent a scandal, and finally Newton inserted into his book a statement that the inverse square law had been deduced from Kepler's third law of planetary motions by himself, and independently by Wren, Hooke and Halley.

This further unpleasantness in respect to scientific work caused Newton much pain and he wanted to suppress Book III, "De Mundi Systemate," containing direct applications of the inverse square law to planetary theory, lunar theory, the theory of comets, tides, etc. Fortunately, Halley's good sense succeeded in preventing such a calamity, and Newton proceeded with the plans originally made. On July 6, 1686, the *imprimatur*, or license to print the book in the name of the Royal Society, was given by the then President, the famous diarist, Samuel Pepys, and within a couple of weeks the work of printing Book I began. Book II reached the Royal Society in March, 1687, and Book III in April of the same year.

THE GRAVITATIONAL DECADE, 1678-1687

Newton's work at the " Principia " was interrupted by an event of political importance. The accession to the throne of King James II was soon followed by an attempt on the part of the King to favour his fellow Catholics in violation of the law of the land. If we can now look with more detached eyes at this exercise of the royal will, and admit that religious tests of any kind are to be condemned, and acts of exclusion against any church are equally undesirable, we cannot treat the King's attempt to force his will upon the universities in the same way. A university is a living monument to knowledge and truth, the degrees and diplomas awarded by a university can be granted only on proof being supplied of learning or research on the part of the candidates for such honours. The only persons qualified to judge the fitness of candidates are the professors of the university : deprive them of this monopoly, infringe in any way upon their free and unfettered decision in academic matters, and the prestige of the honours awarded at their hands is lowered and the university loses in status and dignity as the repository of knowledge.

James had already forced the University of Oxford to accept as a high academic official a man who had no qualification for occupying such a post, but who was a Roman Catholic and as such was a nominee of the King's. In February, 1687, the blow fell upon the University of Cambridge. The particular form that the King's attack took would leave us cold nowadays, since it really amounted to removing restrictions against Catholics as such. But the principle at stake was one that overshadowed the actual issue in the dispute. The Vice-Chancellor of the University declined to carry out the royal mandate to confer the degree of M.A. upon a certain Benedictine monk without his taking the oaths of allegiance and supremacy. Finally, the Vice-Chancellor was summoned to appear before the High Commission Court at Westminster, accompanied by representatives of the Senate of the University. Newton was one of the eight representatives chosen for the purpose, and he maintained a firm determination not to submit and not to accept any compromise.

The infamous Judge Jeffreys presided over the court. The Vice-Chancellor was soon reduced to silence; the other deputies were forbidden to say anything in defence of the attitude of the University, and ordered out of court. "As for you," said Jeffreys to them, "most of you are divines. I will therefore send you home with a text of Scripture, 'Go your way and sin no more, lest a worse thing happen to you.'" The greatest man of the century, not a "divine," was among the Cambridge representatives thus insulted, the man who was just then putting the finishing touches to the book that will outlive scores of royal bullies like James, and foul-mouthed brutes like Jeffreys. The Vice-Chancellor was not only deprived of his office, but also lost his post as Master of Magdalen College, although the University ultimately succeeded in its resistance.

Newton went back to his rooms between the great gate and the chapel at Trinity and proceeded with the proof reading and other tasks in connexion with the publication of the "Principia." Within two months of Newton being ejected ignominiously from the court presided over by Jeffreys, the "Principia" was published, dedicated to the Royal Society " flourishing under his august Majesty James II "!

Newton was then forty-four years of age. He had spent over twenty years in the pursuit of truth, mainly in optics and in gravitation theory. He had remained a bachelor, possibly by inclination; but certainly in accordance with the peculiar traditions that were in force till only half a century ago, and which prevented a college fellow leading the normal life of a human

THE GRAVITATIONAL DECADE, 1678-1687

being. He had remained poor in accordance with the general practice of civilized humanity, that is content to let its men of genius give their all to humanity and receive nothing in return. Newton had reached middle age in a state of semi-obscurity, known to the few like Halley, Wren, Huygens, Leibniz and Hooke, but unrecognized by the masses whose adulation is blindly bestowed and equally blindly withheld.

On the evidence of his relation, assistant and amanuensis, Humphrey Newton of Grantham, who was with Isaac Newton from 1683 to 1689, we know that Newton was "very meek, sedate, and humble, never seemingly angry, of profound thought, his countenance mild, pleasant, and comely." He delivered his lectures to the few who cared to hear them, dictating for half an hour as rapidly as his audience could write; and if nobody came he went back to his work. He took no recreation and had no pastime, devoting all his time to his work. "He ate very sparingly, nay, ofttimes he has forgot to eat at all, so that, going into his chambers, I have found his mess untouched. . . . He very rarely went to bed till two or three of the clock, sometimes not till five or six, lying about four or five hours, especially at spring and fall of the leaf, at which times he used to employ about six weeks in his elaboratory, the fire scarce going out either night or day, he sitting up one night and I another, till he had finished his chemical experiments.

"I cannot say I ever saw him drink either wine, ale, or beer, excepting at meals, and then but very sparingly. He very rarely went to dine in the hall, except on some public days, and then if he has not been minded, would go very carelessly, with shoes down of heels, and his head scarcely combed. . . . He was very curious in his garden, which was never out at order, in which he would at some seldome time take a short walk or two, not enduring to see a weed in it."

This garden was the space between the road and the college on the right-hand side on entering the great gate at Trinity College, and contained his chemical laboratory. "In his chamber he walked so very much that you might have thought him to be educated at Athens among the Aristotelean sect. . . . He kept neither dog nor cat in his chamber." This disposes of the well-known story that Newton made a hole in the door for his cat to get in and out without disturbing him, and then when the cat had kittens, made another and smaller hole for the kittens to get through!

Newton's refusal to go into holy orders is illuminated by the following: "He very seldom went to Chapel, that being the time he chiefly took his repose; and, as for the afternoon, his earnest and indefatigable studies retained him, so that he scarcely knew the house of prayer. Very frequently, on Sundays, he went to St. Mary's Church, especially in the forenoon. . . . As for his private prayers I can say nothing of them; I am apt to believe his intense studies deprived him of the better part." But busy as Newton might be he had time for charitable deeds, and "few went empty handed from him." He helped poor relations and neighbours most generously, tipped the college porters frequently, although he never broke their rest by coming in late at night. He was patient and conscientious—an eminent example of the great worker prosecuting his studies quietly, and little dreaming that his discoveries were to become the glory of mankind.

CHAPTER VIII

THE "PRINCIPIA," 1687

Nec fas est propius mortali attingere divos.

HALLEY

THE "Principia" appeared in July, 1687, in one small quarto volume containing 500 pages. It was illustrated with a large number of diagrams in the form of woodcuts, was issued bound in calf, and was sold at nine shillings per copy. Halley's generosity to and admiration for Newton are both apparent from the following, taken from a letter written by the former on July 5, 1687: "I have at length brought your book to an end, and hope it will please you. The last errata came just in time to be inserted. I will present from you the book you desire to the Royal Society, Mr. Boyle, Mr. Paget, Mr. Flamsteed, and if there be any else in town that you design to gratify that way; and I have sent you to bestow on your friends in the University 20 copies, which I entreat you to accept. . . . I hope you will not repent you of the pains you have taken in so laudable a piece, so much to your own and the nation's credit, but rather, after you shall have a little diverted yourself with other studies, that you will resume those contemplations wherein you had so great success, and attempt the perfection of lunar theory, which will be of prodigious use in navigation, as well as of profound and public speculation."

In view of King James II having been Lord High

Admiral early in his brother's reign and the probable value of Newton's work in navigation, a copy of the " Principia " was presented to the King by Halley. Did James make much use of it in his many voyages across the sea after his indignant subjects had driven him out of the country sixteen months later ?

The great classics of science are rarely read in subsequent generations. Whereas the young school boy or girl is introduced to Milton's work in the original, and learns, if judiciously taught, to admire and perhaps to emulate the great lines of our incomparable national poet, no boy or girl at school ever sees a copy of the book written by his junior contemporary, Newton. We do not blame modern educationists for preferring modern books on mathematics and science as text-books for modern pupils. The continued progress of science forces us to modernize, to condense and to systematize the mathematical and scientific knowledge of the past in a manner suitable for the consumption of present-day learners. Yet one would wish that a little time could be found to introduce the pupils of our schools, and especially of our universities, to the actual words of the great original thinkers in science. If scientific results are impersonal and objective, scientific research embodies the mind, the soul, of the worker, and the essence of education is surely the bringing to bear upon one mind the influence of another mind, a forerunner in the search for truth.

The " Principia " was written and issued in Latin, the international language of learning till a few generations ago. Newton was not only a discoverer of unique power, but also an elegant exponent of his views. Said Laplace, the great scientific representative of a different and of a hostile nation : " . . . tout cela présenté avec beaucoup d'élégance, assure à l'ouvrage des Principes, la prééminence sur les autres

THE "PRINCIPIA," 1687

productions de l'esprit humain"—extreme but not exaggerated praise.

The "Principia" is not a compendium of knowledge reproducing the learning of the past with original touches here and there. It is one of the most consistently original books ever written. The fundamental laws of mechanics are here formulated for the first time, the mathematical ideas required in the arguments are those invented by the author himself, the forces invoked to explain the motions considered are those arrived at by the author's own intuition, the phenomena explained had never been satisfactorily accounted for before, the problems initiated have since occupied the best minds of humanity for two centuries and a half, the results deduced have been the admiration of successive generations of astronomers and mathematicians—and pervading the book there is a spirit of dignity and simplicity, a sanity of outlook, a soberness of expression, and a grandeur of conception, that place both the author and his work on an unequalled level of eminence. Immortals themselves, Halley and Laplace could appraise at its true value genius superior to their own.

It is quite impossible to convey an adequate impression of Newton's achievements in the "Principia" in a brief space and in non-technical language. A mere enumeration of the topics discussed is sufficient to impress with wonder and admiration one so versed in the subjects of mechanics and astronomy as to appreciate the importance of these topics. First there is an introduction, in which, after some definitions and the consideration of the meaning of space and time, Newton formulates the three laws of motion that are still called by his name, with deductions and applications of interest and significance.

We have already stated how significant for the development of dynamical astronomy were the dynamical experiments of Galileo; without them the whole subject was quite unapproachable. Newton therefore gives in the " Principia " a clear statement of the laws of motion deduced from these experiments. The first two laws have already been mentioned, namely, that the absence of force implies uniform motion in a straight line, while change of motion or acceleration is determined by force. A third law must now be added: its importance in connexion with our present subject is hardly ever explained adequately, whereas without it Newton's work would have been wofully incomplete.

The third law of motion is the famous statement that action and reaction between two bodies are always equal and opposite. We can readily see the importance of this law in Newton's train of thought. Newton's use of the principles of dynamics was in itself a lesson in their clarification and of their significance.

Astronomical progress from the flat earth to the spherical earth and from the fixed spherical earth to the earth as a rotating and revolving planet, implied more than this bald statement seems to convey. The human intellect needs an anchorage in space and time as much as it needs one in its spiritual life. This was found in the flat earth. Man was banished from the flat earth and sent to wander in bewilderment round the surface of a sphere. Still, he had the centre of the earth, the fixed centre of all things. Copernicus completely banished man's intellectual anchorage from the earth and sent it to the sun. Humanity struggled against the decree of banishment, but finally and with a bad grace it submitted, and promptly appropriated the centre of the sun as the fixed point, the hub of the universe, the centre of all creation. Kepler's discovery of the more accurate paths of the planets left this state of affairs unchanged—

THE "PRINCIPIA," 1687

the sun remained enthroned in mighty splendour at the centre of the universe, directing his vassals, the planets, and their satellites.

Meanwhile Galileo discovered the laws of dynamics. When we discuss motion, especially variations in motion, we necessarily look for a comparison point. When we say that a body is at rest we cannot mean anything else than that it is not changing its position relative to a set of bodies which we take to be the standard. Is the reader at rest while reading this page? Is he reading this book in his house seated quietly near the fire—or comfortably in bed? He is clearly at rest—yet he is being dragged round with the earth's diurnal rotation and with the earth's annual revolution. What then shall be our standard of rest in dynamics? To which body are we to refer motions in order to decide whether they are uniform or varying, and what the variations are?

It is a remarkable feature of natural phenomena that in spite of their apparent complexity, Nature always contrives to make our researches into her secrets amenable to our limited intellectual powers. Galileo carried out his experiments on the earth—he ignored the earth's motion completely. The effects of the earth's motion on his experiments were negligible, because the earth's angular rates of rotation and revolution are both so small as to produce no appreciable effect on ordinary terrestrial dynamical phenomena. These and other considerations made it possible for Galileo to arrive at correct conclusions about the earth's gravity, and about the effects of forces on bodies. Would we even now be in possession of reasonable views on dynamics if the earth were a very small body, say ten miles in diameter, or if it rotated on its axis once in ten minutes, or revolved round the sun once in twenty-four hours? The phenomena of solar and terrestrial gravitation would be so confounded together, as well as with variations in the earth's gravity itself, that no rhyme or reason could have been discovered in the subject by finite intelligences. Let us be grateful to a kindly Nature for presenting her problems to us step by step, and not confounding us with a too precipitate confrontation of all her complications.

Galileo could take the earth as his dynamical standard; in fact, the surface of the earth wherever he happened to be was good enough for his purposes. This would of course be absurd in dealing with extraterrestrial dynamics. People still take the sun as dynamical standard, and we often hear loose talk about the sun being fixed at the centre of the solar system. But Newton had to postulate universal gravitation, that all bodies attract one another—the sun pulls the planets and the planets pull the sun, the earth pulls the moon and the moon pulls the earth; it were therefore ridiculous to imagine the sun "nailed down" and not yielding at all to the attractions of all the other bodies in the solar system. For this problem Newton used the third law of motion. Let the sun and earth attract one another, but let the force of attraction exerted by each on the other be the same, yet in opposite directions. The accelerations produced will be inversely as the masses, and it is at once seen that the point which divides the distance between the centre inversely as the masses will in fact have no acceleration at all! A little mathematical reasoning enabled Newton to extend this to the whole solar system; the centre of gravity of the whole solar system has no acceleration.

What about the so-called fixed stars? Do they attract? Why not? Of course they do. Then there are external forces on the bodies of the solar system, and its centre of gravity has not zero acceleration. We have here once more an example of the benignity of Nature towards human scientific research. The stars are indeed so far off that Newton made no serious error in neglecting the accelerative effect of their gravitational pulls. He had no actual measures of the distances of the stars, but the intuition of genius stood him in good stead, and an estimate of stellar distances that he made is remarkably like those obtained by direct measurement a century later.

Newton thus took the centre of gravity of the solar system to have no acceleration. The logical thing would be to say that it has a uniform speed in a constant direction. But Newton had no means of discussing the motion of the solar system as a whole—this came over a century later. He therefore took a deep plunge and as we shall see later he made the centre of gravity be at rest. It was left to the nineteenth century to correct this, and to the twentieth century to draw far-reaching conclusions from the correction.

One of the most interesting features of the "Principia" is this continual provocation of thought on the part of the reader. The main theorems and propositions are given in clear, brief, crisp and conventional language. The implications of the theorems are discussed in corollaries, and particularly in the "Scholia." A scholium is a general remark or conclusion that Newton often appends to some particularly important proposition and set of propositions. The scholia contain a vast amount of stimulating discussion, and they shed a large amount of light upon the whole subject, by means of the citation of the authorities, experimental investigations and results, and anything else that helps to illuminate the matter in hand.

The main work is in the two books of the "De Motu Corporum" and in a third book on "De Mundi Systemate." Book I (the first part of "De Motu") commences with a brief account of the principles of the method of fluxions. We have already seen that Newton had written brief treatises in fluxions about twenty years before. His discoveries in this field were communicated in manuscript to several mathematicians, and some correspondence with Leibniz had taken place as far back as 1676. Eight years later Leibniz published in the "Acta Eruditorum" —a mathematical journal edited by himself at Leipzig in Saxony—an account of the differential calculus, essentially the same method as fluxions, but with different, and in fact more suitable notation, giving an oblique reference to Newton's work of a similar kind. Leibniz developed the subject further in another paper two years later, 1686. No doubt Newton read these papers and felt that the time had come when he should give to the world his own theory of fluxions. He accordingly incorporated in the "Principia" a section on this subject, in the form of eleven lemmas or preliminary propositions required for the sequel. The treatment is very brief, and even at this late hour—twenty-one years after his developing the method—the notation is not given; this was not printed till six or seven years later in the works of John Wallis! Meanwhile Leibniz' method and notation were being used on the Continent with phenomenal success for solving difficult and important problems.

After this preliminary, Newton attacks the motion of a body under an attraction directed to a fixed point. When the path is a conic section the force must bear a certain relation to the distance—this is discovered, and then particularized for the case where the centre of attraction is at a focus of the conic, when the law is that of the inverse square of the distance. Detailed developments of this type of motion follow, and Newton examines the way in which any particular orbit is described, showing how to find the position at any time.

This would suffice for the discussion of the planetary motions in accordance with Kepler's third law. But Newton did not merely attribute gravitational properties to the sun as acting on the planets or to the earth as acting on the moon. It soon became clear to him, especially on the basis of his third law of motion, that every heavenly body attracts every other heavenly body, that our earth attracts and is attracted by the moon, that the sun attracts and is attracted by the planets, that Jupiter attracts and is attracted by his satellites. It is a further instance of the encouragement that Nature gives to scientific discovery that the sun is so large compared to the planets, containing about a third of a million times as much matter as the earth, indeed about eight hundred times as much matter as all the planets put together, and further that the planets are very far apart. The result is that it is permissible in discussing the motion of any one planet to omit the pulls exerted upon it by the other bodies in the solar system, and to assume that it moves under the sun's influence alone. The same applies to a satellite. Satellites are usually so close to the planets round which they revolve, that even the sun's disturbance can be neglected, at least as a first approximation. Again we might ask ourselves what progress in astronomy would have been possible if, say, the planets had masses which were considerable fractions of the sun's mass, or if two planets came occasionally very close together, or if our moon were very much further away from the earth than it actually is.

The problem of the sun and one planet, of a planet and one satellite, was the problem discussed by Newton primarily. But he had to go to a second approximation in order to explain a number of irregularities observed in the motion of the moon, and known for centuries. Newton's law of universal gravitation suggested what had not yet been discovered, namely, that even Kepler's laws were only approximately correct. Happily, Kepler lived at a time when these mutual disturbances or perturbations of planets were not measurable—instrumental accuracy had not risen to this level. He could therefore deduce the elliptic motion round a focus without being troubled by the really considerable deviations therefrom.

In the case of the moon the disturbances were so great as to have been observable long before Tycho Brahe, Kepler and Newton. The moon thus presented an excellent means of testing not only the first rough results of Newton's theory, but also its more accurate consequences. The preparation for this is contained in the next section of Book I, where Newton discusses the effects of any law of force, the case of paths which are not closed, so that the moving body does not return to its original position exactly.

A very important and fundamental theorem now follows. Newton was not content to attribute gravitation to the sun or to any other body as a whole. He sought to discover the gravitational effect of any body on the basis of allowing each particle of the body to attract for itself, in which case, of course, it cannot be guessed off-hand what the total effect would be, even in such a simple case as a sphere—the really important case in astronomy. In a singularly beautiful manner Newton proves that if the matter of a sphere is arranged symmetrically round the centre it produces at any outside point the same force as if all the mass of the sphere were concentrated at the centre of the sphere. It is believed by some authorities that Newton's delay in publishing the gravitational discoveries was due to the difficulty he experienced in proving this theorem.

What was the use of this theorem? Could not Newton say that this is the law which he postulates, since heavenly bodies are spheres? Newton knew, however, that in this consideration of the attraction produced by the separate particles of a body he had the key to many mysteries—the flattened shape of the earth, precession, tides, etc. Some theorems on the attractions produced by non-spherical bodies follow, and Book I comes to an end with a somewhat irrelevant discussion of motions that refer to Newton's conception of light as due to tiny corpuscles moving with great speed—this last piece of work is now entirely out of date and discarded.

Book II of "De Motu Corporum" deals with the problem of motion in a medium of some kind which might produce a resistive effect to the motion. Newton takes this resistance to be proportional first to the velocity, and then to the square of the velocity; finally he takes the resistance to be in two parts, one proportional to the velocity and the other to the square of the velocity—thus anticipating in remarkable manner modern ideas on air resistance in aeronautics. The results are then applied to motions under central attractions where resistance of this kind exists, and interesting results in the form of spiral motions deduced.

This leads Newton to the consideration of fluids in general and their equilibrium under gravitational attractions—a subject which has become of great importance in connexion with the discussion of the shapes of heavenly bodies. Since the pendulum is important in measuring the force of gravity at the earth's surface, Newton examines the motion of a pendulum in a resisting medium like air, and naturally proceeds to discuss the way in which the resistance depends upon the size and shape of the bob of the pendulum, upon the density of the medium, upon the velocity, etc. The study of wave motion in fluids comes next, and the velocity of propagation of such a wave is obtained by Newton, with obvious application to the propagation of sound through air. Experimental verification was attempted in that famous walk in the cloisters of Neville's Court at Trinity, where the modern don or modern undergraduate never fails to test the quadruple echo which Newton claims to have heard there. Experimenting in this walk Newton measured the velocity of sound roughly, with qualitative if not quantitative verification of his mathematics.

One further point had to be settled before Newton could proceed to the application of his theory to the solar system. The Cartesian theory of vortices was then the accepted view of the cause of the motions of the planets. This theory had to be demolished first, and Newton disposed of it in a few simple propositions. He first shows that if the vortices exist and persist then they must be such that the fluid in each vortex rotates as a whole, as if it were one rigid body, without differential angular motions. It follows that the further a point is from the axis of the vortex, the faster it moves, so that the vortex theory gives velocities to the planets which are in complete contradiction to the observed velocities, the planets being known to move faster when nearer the sun and slower when farther away from the sun. "Hence the Vortex Theory is in complete conflict with astronomical facts, and so far from explaining celestial motions would tend to upset them."

The stage is now set for the final scene, the reconstruction of the phenomena of the solar system by means of the law of universal gravitation. Book III, "De Mundi Systemate," the book that Newton was for suppressing when worried about Hooke's claims to prior discovery, shows us Newton at the very pinnacle of his intellectual achievement. Here we see him as if he were a creator conjuring up before our charmed vision whirling and rotating planets and satellites, flowing and ebbing tides, erratic wanderings of comets.

A brief preface in the form of "Hypotheses" contains two statements that throw a brilliant light upon the scientific character of Newton's thought. Right through the Middle Ages all thinkers differentiated somehow between earthly and celestial phenomena, postulating mystical causes and effects which bore little if any relation to ascertained facts. Newton broke with this most forcibly. "Like effects in nature are produced by like causes," he asserted, "as breathing in man and in beast, the fall of stones in Europe and in America, the light of the kitchen fire and of the sun, the reflection of light on the earth and on the planets." Thus was humanity finally redeemed from the bondage of many centuries to belief in the perfection of the heavenly, imperfection of the earthly, and all phenomena were reduced, or raised, to the same level of rational causality.

It is interesting that Newton takes as his first proof of the inverse square law the fact that the periods and distances of the satellites of Jupiter and Saturn follow Kepler's third law, i.e. the squares of the periods are proportional to the cubes of the mean distances. Why does not Newton give the case of the moon first? We have here another illustration, perhaps, of Newton's preference for the immediately visible. Anybody can verify the relationship between the periods and distances of the four large satellites of Jupiter—not much observation is needed; whereas the case of the planets round the sun needs careful measurement, and the case of the moon is a somewhat recondite application of dynamical principles, not well understood by many men in Newton's day.

Having established the gravitation of Jupiter, Saturn, the sun and the earth, Newton then leads on to the grand generalization of the law of universal gravitation, the force between two bodies being proportional to the product of their masses, and inversely proportional to the square of the distance between them. He shows that no appreciable resistance is experienced by the planets so that their motions may be expected to go on indefinitely. He then determines the motion : the centre of gravity of the sun and all the planets is at rest—a great improvement on the Copernican system in which the centre of the sun was taken to be at rest—the sun never receding much from this point, while the planets move round the sun in ellipses, with a focus of each ellipse at the sun.

Newton establishes the uniform rotations of the planets by reference to the moon's librations, already mentioned here, and then proceeds to discuss the shape that a planet should assume, using the knowledge of the apparent flattening of Jupiter at its poles, as told him by Cassini of Paris and Flamsteed of Greenwich. He proves that the rotation of a planet will cause such flattening, and discusses pendulum observations in the case of the earth to show that the variation in the earth's gravitation from place to place on its surface is explainable by this flattening.

The flattening of the earth explains the precession of the equinoxes, a phenomenon known since the middle of the second century B.C., and completely misunderstood till Newton gave it its natural explanation. Since the earth is not an exact sphere, the attractions exerted upon it by the sun and by the moon do not pass through the earth's centre, but meet the earth's axis at some little distance away from the centre. The effect in each case is to produce a turning effect, or moment, about an axis lying in the equatorial plane of the earth; this gives a gyroscopic or top motion, in which the axis of the earth describes a cone in the heavens. This is one of the most beautiful deductions from Newton's theory that can be imagined and the agreement quantitatively leaves little to be desired.

Having shown in considerable detail that the irregularities observed in the motions of the moon and of the satellites of Jupiter and of Saturn (the only secondary bodies then known) can be explained by

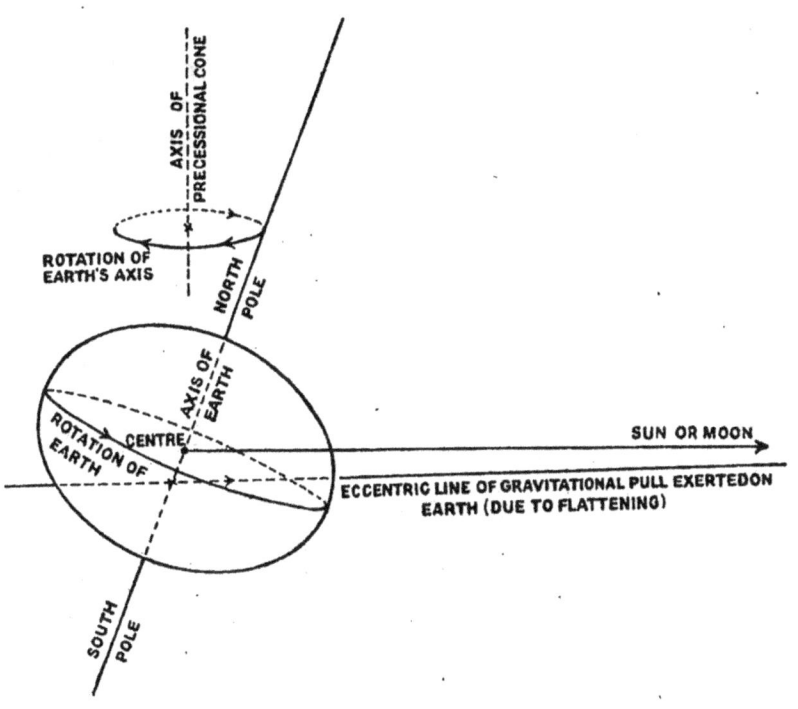

Fig. 8

CAUSE OF PRECESSION

Pull of sun or moon on earth does not (owing to flattening of the earth) pass through its centre, as shown here. A torque is produced about an axis through the earth's centre perpendicular to the plane of the paper. This causes the earth's axis to "precess," like a top, in the reverse sense to the earth's own rotation.

gravitation as due to the perturbative effect of the sun, Newton proceeds to the question of tides. He at once attributes the tides to the differential gravitative effects of the sun and moon. If we consider the earth and the moon, the former containing on its surface a flowing substance like water, we see that the attraction produced by the moon on unit mass of the water on the side of the earth away from the moon must be less than on unit mass on the side towards the moon, while on the earth itself the attraction corresponds to the total mass of the earth concentrated at the centre of the earth, and thus per unit mass produces effects lying intermediate between those on the further and nearer waters. The result is that the nearer waters are dragged towards the moon more than the earth as a

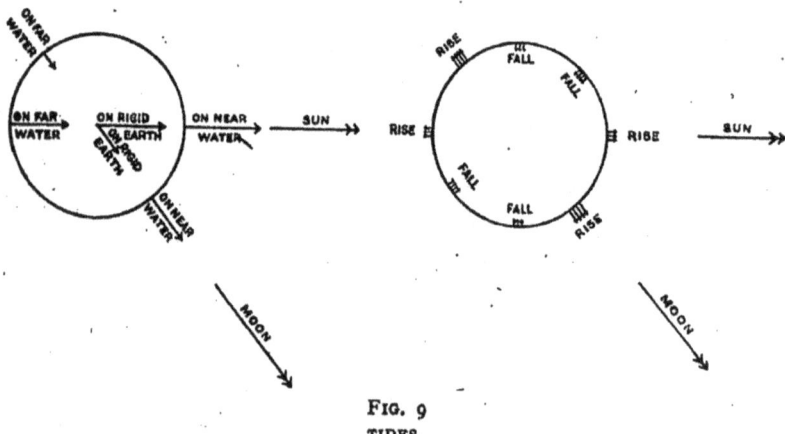

FIG. 9
TIDES

Gravitational pulls of Sun and Moon on Earth, and on waters on near side and on far side of Earth.   Tides due to sun and moon respectively.

whole, and the further waters are dragged towards the moon less than the earth as a whole. The waters are thus heaped up on the two sides of the earth, towards the moon and away from the moon.

The same applies to the sun. But the moon is so near that its tidal effect is greater than that of the sun; hence the tides follow the moon as a whole. Hence, also, we get the very high tides at new and full moon, when the sun and moon reinforce one another's tidal

THE "PRINCIPIA," 1687

effects, and low tides at the quarters, when the sun and moon tend to neutralize one another's tidal effects.

Finally we get a rational and beautiful treatment of comets. Newton's knowledge of comets was somewhat scanty when he discussed them with Flamsteed several years before he wrote this theory. But in the "De Mundi," he produces a remarkable account of the subject. He proves that comets must move in

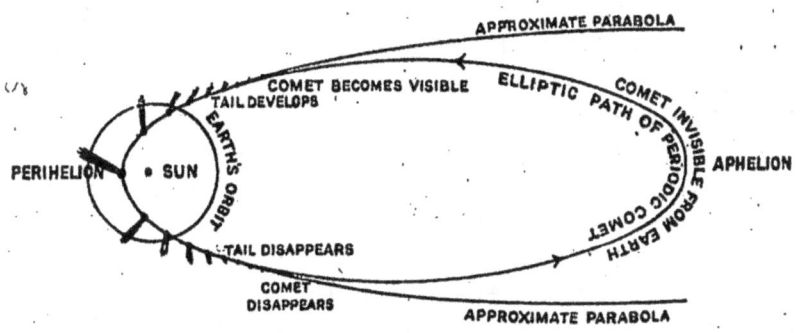

FIG. 10
PATH OF PERIODIC COMET
Long ellipse with one focus at the centre of the sun.

interplanetary space and shows that they also move in conics about a focus. Kepler's ellipses were one type of conic—the cometary paths are another type, vindicating wonderfully the generality of Newton's gravitational ideas. Newton then reverts to the parabola as an approximation to a long cometary ellipse, and discusses how to find the approximate parabolic path of a comet from three observations of its position as seen from the earth—a result of fundamental importance, since by its means one can deduce the future path of a comet quickly after its first appearance, so as to facilitate observation and a more accurate determination of its path subsequently.

There is much in the " Principia " to which we have not referred and which is of great importance in the theory of the heavenly motions. We have talked about the masses of the planets and of the sun. How can we find these masses ? Nothing is simpler when we have once grasped the fundamental idea of gravitation. If a planet has a satellite, then by the motion of the satellite we can calculate the gravitative effect of the planet at the distance of the satellite. But we have the gravitative effect of the sun for all sorts of distances. By comparison we get the ratio of the gravitative effects of the sun and of the planet for the same distance, and this is the ratio of their masses. In this way Newton could find the ratio of the mass of the sun to that of the earth or of Jupiter. Now we can find this ratio for all the planets in this way, except Mercury and Venus which seem to have no moons.

No limit can indeed be foreseen to the developments consequent upon Newton's work. The " Principia " is not an ordinary book which plays its part in the progress of knowledge and then disappears. Its theorems and methods, principles and laws, are pronouncements that humanity has studied for generations. Over and over again the greatest triumphs of human thought have been achieved as the direct consequence of Newton's genius. The exact working out of the motions of the bodies in the solar system at the hands of Laplace, Lagrange and their successors has introduced into astronomical prediction such a degree of precision that popular concern is aroused when an eclipse takes place a couple of seconds earlier or later than the time calculated in advance, and messages are cabled round the world to announce this fact. The slight differences between the observed motion of Uranus and that calculated on the basis of the sun's attraction and the perturbations due to the known planets, led to the discovery on paper by mathematical calculation of the till then unknown planet Neptune. The difference between the motion of Mercury and that deduced from theory has been one of the main planks in Einstein's platform. And what is perhaps the outstanding event in the science of our own generation—the theory of relativity—is a wonderful vindication of the sagacity of the originator of universal gravitation, for it is the universality of gravitation that is the chief argument in Einstein's view, and the universality of gravitation is the distinctive characteristic of Newton's conception of this force.

But one of the most historic vindications of Newton's genius was coupled with that of none other than his friend and admirer Halley. Man is always impressed with the power of foretelling events yet to come, and Halley was able to present to the world a remarkable piece of prophecy, the fulfilment of which was left to the generations that came after him and made him one of the immortals of astronomy.

A great comet had appeared in 1682, and Halley had observed it carefully, deducing its path when near the sun. When Newton published the "Principia" and demonstrated that long elliptic paths with a focus at the sun are possible, he concluded that what appears to be a parabolic path of a comet may really be a long elliptic path, and suggested that the periodic time of a comet moving in a long ellipse could be ascertained by its return after a long interval of time. But a comet becomes visible to us only when it is near the sun, whose effect is to eject matter from the cometary nucleus in the form of one or more tails. Thus a comet is invisible except for a very short time when it is near the sun, and a comet could not be traced all round its elliptic orbit in order to verify Newton's theory.

The sagacity of Halley came to Newton's help.

Halley set himself to examine all reliable observations of comets that had appeared in the past, and where sufficient material existed he calculated the parabolic path which Newton took to be the first approximation. He discovered that in the year 1607 a comet had been observed whose approximate parabolic path was very nearly identical with that of 1682. This was interesting as a coincidence, perhaps. But Halley found that in the year 1531 a comet had been observed whose approximate parabolic path was also nearly identical with that of 1682—further, the interval 1531–1607 is nearly the same as the interval 1607–1682. Halley became convinced that this was not a cometary accident or coincidence, but rather a cometary habit, and concluded that he had indeed discovered a periodic comet, the existence of which was suggested by Newton. Any deviations from exact identity of the calculated parabolas, and any inequality between the two intervals, Halley attributed to planetary perturbations.

Newton encouraged Halley in this view, and in 1705 Halley announced that the comet he had observed in 1682 would reappear—after suffering severe perturbation owing to a very near approach to Jupiter—about the end of 1758 or the beginning of 1759. Halley had looked back into the centuries and observed the similarities of the cometary paths of 1531, 1607 and 1682; he looked forward, with the aid of Newton, into generations to come and saw his comet wending its long and lonely way, past the orbit of the earth, past the orbit of Mars, past Jupiter, past Saturn, deeper and deeper into space. Halley and Newton grew older and older. Newton passed away and Halley was left to continue alone the dynamical watch of the comet. He rewrote his prediction, and, feeling that he would never see the year of the comet's return, expressed the hope that if the comet did come back as predicted "impartial posterity will not refuse to acknowledge that this was first discovered by an Englishman."

Within a month of the time given by calculations based upon Newton's laws the comet reappeared. Halley was dead, but grateful posterity will forget neither Halley's personality nor his nationality. Halley's Comet reappeared in 1835 and in 1910, in better and better agreement with calculated time and place—a lasting monument (not permanent since comets disintegrate, and Halley's Comet is believed by some to be showing signs of decay already) to the memory of Newton and Halley. It is exceedingly appropriate that Halley's memory should have thus become linked with that of the man whom he admired so selflessly and encouraged with such devotion.

Such are the contents of the " Principia " as issued in 1687, written by the man driven out of his court by Jeffreys and admonished to mend his ways, and printed at the personal expense and risk of the man who could best appreciate its contents.

CHAPTER IX

THE TRANSITIONAL DECADE, 1687-1696

> Has matter more than motion ? Has it thought,
> Judgment and genius ? Has it framed such laws,
> Which, but to guess, a Newton made immortal ?
> EDWARD YOUNG : " On the Being of a God "

EVERY new idea has to overcome prejudice and mental inertia. The publication of the " Principia " was a turning point in the intellectual history of mankind ; yet Newton came on the scene when Descartes' vortex theory had become well entrenched, enjoying the prestige associated with the immortal name of this philosopher and mathematician. On the Continent, in fact, Newton's theory made little progress for some time. The supposed suggestion of action at a distance, without any hypothesis as to how the action was propagated in the intervening space, caused considerable opposition on the part of Leibniz and others, especially when personal jealousies confused the issue. It will already be clear to the reader that Newton was the last man in the world to postulate non-physical or anti-physical conceptions ; he did indeed make an attempt at a physical explanation of gravitation based upon the ether. But this effort produced nothing really useful, and more than two centuries had to elapse before a satisfactory view of gravitation began to emerge.

Things were better in England, where Newton's work found ready approval among those few whose opinions were worthy of consideration. Within three years of the publication of the "Principia" its contents were taught at Cambridge, St. Andrews and Edinburgh, while Oxford followed suit before very long. The "Principia" helped to establish mathematical studies on a firm basis, especially at Cambridge. Non-mathematical thinkers took special pains to study the physical portions of the book, and the influence of Newton's ideas spread quickly among the best minds of the day. The "Principia" sold quickly, and one hopes that Halley did not lose on his generous transaction. Copies of the first edition became scarce, and people paid four or five times the original cost for a copy. One case is known of a Scotsman who wrote out the whole book by hand, as he could not get a copy at a reasonable price.

Following Halley's advice Newton took a rest from scientific work. In November, 1688, William of Orange landed at Torbay and marched upon London. James II fled the country and William was invited to share the throne with his wife, James' daughter, Mary II. In January, 1689, the so-called Convention Parliament was summoned to Westminster, and, as Macaulay says : "Among the crowd of silent members appeared the majestic forehead and pensive face of Isaac Newton . . . he sate there, in his modest greatness, the unobtrusive but unflinching friend of civil and religious freedom." Parliament did not hear Newton's voice in debate during the thirteen months that he served as M.P. for Cambridge University—till the dissolution in February, 1690—yet Newton's presence was not unfelt. He was a keen supporter of the new royal house and a Whig in politics, being a strong believer in religious toleration, and a stout opponent of all forms of oppression—in particular of the oppression, religious and otherwise, practised by the Stuarts. He did much to make and keep Cambridge loyal to King William and Queen Mary.

Newton's presence in London served to accentuate the unsatisfactory nature of his own personal position. He was nearly fifty years of age, and he had spent all his manhood days in the most unremitting scientific toil. No recognition of a tangible nature had come to him. He was still college fellow and university professor, conscious of a harrowing poverty, which narrowed his existence and prevented him from satisfying the needs of his relations and other dependents, and which had necessitated his acceptance of Halley's offer to shoulder the financial responsibility involved in printing the " Principia."

Locke, Pepys and other friends of Newton who occupied prominent positions in public life tried to secure for him some public post worthy of his greatness. In particular, Charles Montague, an intimate friend of Newton's and a fellow-member of the Convention Parliament, who was rising quickly in the political firmament, did all he could for him. But the great men of the day were too busy to bother about a mathematician whose work they could not understand. When the post of Provost (or Head) of King's College, Cambridge, fell vacant, Newton was nominated, but ultimately failed to secure appointment because he was neither a Fellow of King's nor in priest's orders. The dissolution saw Newton back in Cambridge, resuming his bachelor life in a college cell.

Meanwhile other things had happened to depress Newton. During the year of his membership of the House of Commons his mother, Mrs. Smith, became seriously ill, and Newton left everything to minister to her. He was a skilful nurse ; he sat up with his mother whole nights, and did everything in his power to lighten her sufferings. She died without having enjoyed the satisfaction of seeing her son Isaac receive the token of his nation's gratitude that was his due.

THE TRANSITIONAL DECADE, 1687–1696 125

Did his mother's death turn Newton's attention to theology ? As early as 1690 his interest in theological discussion began to manifest itself and was encouraged by others. He corresponded with various people about gravitation, and drew up for John Wallis in 1692 a few propositions on fluxions, which were printed in the second volume of Wallis' works ; he corresponded with Leibniz about his mathematical work ; he gave free rein to his lifelong interest in alchemy and studied many old and contemporary books on this subject ; he interested himself in a method due to Boyle for making gold out of mercury ; but one of the most important of Newton's productions during the period 1690–1693 was in the domain of theology.

In 1690 Newton wrote, but did not publish, a discussion of "Two Notable Corruptions of the Scriptures, in a Letter to a Friend." These referred to 1 John v, 7, and 1 Tim. iii, 16, and Newton claimed that the trinitarian interpretations of these passages were not justified by the correct original text. It is held by many that Newton did not believe in the Trinity; at any rate, it is clear that he did not accept all the tenets of the prevailing Church unquestioningly, and was as independent in theological belief as in scientific investigation. Newton also wrote then and later on the Prophecies of Daniel, on the Apocalypse of St. John, and about many other religious topics. But the most famous of his theological writings are his letters to Bentley.

Richard Bentley was a brilliant young scholar who afterwards became Master of Trinity. He was interested in gravitation, and Newton gave him advice on the preliminary reading in mathematics needed in order to be able to follow the "Principia." In December, 1691, the famous Robert Boyle died and left some money to endow a lectureship in the interests of Christian orthodoxy. Bentley was appointed the first lecturer under the bequest, and he decided to devote two out of the eight lectures to a consideration of divine providence as evidenced by the physical universe. But he encountered some difficulties and applied to Newton for help. Newton replied in four letters, written between December, 1692, and February, 1693.

Bentley's chief difficulty was the argument used by the Roman poet Lucretius, who had claimed that there is no need for postulating a divine creation, since matter evenly scattered throughout space, and endowed with the power of gravity, would form the universe as we know it. Newton replied that matter in a state of diffusion could not, if left to itself, and without the aid of Divine will and design, produce differently constituted bodies like the bright sun and the dark planets, and could not endow the planets with their characteristic nearly circular motions round the sun. Plato's idea that the planets originated very far away and then fell inwards till they reached their present orbits, was proved by Newton to be inconsistent with the law of gravitation, since it would imply a sudden halving of the sun's attractive power when the planets assumed their orbital motions. Nowadays we would not take it as obvious that diffused matter, or nebula, will not explain the formation of the solar system—a nebular hypothesis has in fact become associated with such names as Kant, Herschel and Laplace.

Meanwhile Newton's bachelor habits began to tell upon his health. He must have had an excellent constitution, since he enjoyed a long life in spite of his many years of irregular and insufficient sleep, and negligence as regards food and other bodily comforts. He became physically unwell and suffered from nervous irritability. He began to brood over the

THE TRANSITIONAL DECADE, 1687-1696

injustice done him, which his experience as a member of the House of Commons had made him feel most keenly. Public offices of honour and of profit were being bestowed upon the deserving and the undeserving, men whose merits were incomparably lower than his own were playing a prominent rôle in the life of England; while he saw no prospect for himself but the obscure and desolate life of the college recluse. Always difficult in personal relationships, Newton did not hesitate to charge his friends with being false to him, and wrote some silly letters to this effect.

Newton's behaviour and unhappiness, aggravated as he himself testified " by sleeping too often by my fire [1692] . . . so that when I wrote to you [Locke] I had not slept an hour a night for a fortnight together, and for five days together not a wink," alarmed his friends very much. Highly coloured rumours were spread about his health. On the Continent, where Newton's work in optics had made a considerable impression, and his gravitational work was slowly penetrating, rumours were spread that he had become insane, and an interesting story was told to account for it. It was said that a little dog called Diamond had caused the loss by fire of Newton's laboratory, together with a manuscript containing the results of many years' labours. To add verisimilitude to the narrative the story went on to say that the meek and mild Newton " rebuked the author of it with an exclamation, ' O Diamond! Diamond! thou little knowest the mischief done!' without adding a single stripe." We have already seen that Newton did not keep any domestic animals. The fire, of which so much was made, and which produced only little damage, took place much earlier, before Newton had even begun to write the " Principia."

In 1691 Newton resumed his work on the motions of the moon. This was a most difficult subject—lunar theory is even now the most difficult branch of dynamical astronomy—and no doubt Newton's nervous excitability was aggravated by these head-racking studies. He desired to get the best possible observations of the moon in order to verify the results of his mathematical inquiries, and tried to get into touch with the Astronomer Royal, Flamsteed. But it was not till 1694 that really useful information was forthcoming, and Newton spent much of his time on his favourite chemical experiments. When Flamsteed was ready a correspondence ensued between him and Newton which lasted nearly two years. Flamsteed gave Newton observations of the moon, as well as the values he had deduced for the refraction produced by the air—a matter of the greatest importance in connexion with the observations of the exact positions of heavenly bodies, since the apparent positions are always affected by the bending of the light rays in coming from outside the atmosphere into the atmosphere of the earth before reaching our eyes or our telescopes. Newton was able to make considerable progress with the theory of both subjects—the moon and refraction—and was very grateful indeed. Yet personal unpleasantness intervened for which Newton himself must bear a portion of the blame.

Flamsteed was a very peculiar person, a permanent invalid who was never free from headaches and other more serious physical affections: he was not treated at all generously as Astronomer Royal, and had to pay his way by acting as a vicar at the same time as doing his astronomical work. He was a religious man and had an antipathy to Halley—who was not very orthodox in his belief, and who had offended Flamsteed in astronomical matters. Newton was a very kind man, but he had gone through considerable physical suffering and had become impatient and excitable. He always became irritable in personal matters: in

THE TRANSITIONAL DECADE, 1687-1696 129

the words of his friend Locke: "Newton was a nice * man to deal with, and a little too apt to raise in himself suspicions where there is no ground." Further, Newton was a particular friend of Halley's who had done so much for him.

This triangle led to difficulties and many hard words were said between Newton and Flamsteed, with Halley as the provocation. It is sometimes not uninteresting to note the pettinesses of the great—a link is thus forged between more humble mortals and the men of transcendent genius, who would otherwise lose all semblance of frail humanity.

While Newton was thus betraying his human weaknesses against the greater weaknesses of Flamsteed, a complete change in his life and fortunes was preparing for him. His friend Charles Montague had not been false to him, but had used the first available opportunity to give Newton evidence of his friendship, and at the same time save the reputation of England in regard to her treatment of the greatest Englishman of the day. Newton was busy with chemical experiments once again, and the call that now came to him was one in which he could exercise to the full his lifelong interest in metals and alloys.

---

* Not in the modern sense, but in the Shakespearean sense of "fussy," "*difficile.*"

9

CHAPTER X

THE GUARDIAN OF THE NATION'S COINAGE, 1696–1727

"Thy silver is become dross."
Isaiah, i, 22

NEWTON had spent thirty-one years as a college graduate and fellow and university professor. This was just half of his adult years; he had thirty-one years more to live, and these he spent in the midst of the public affairs of the nation and as the doyen of British science.

From the time of Queen Elizabeth England had had a coinage based upon bimetallism—both gold and silver were legal tender. But the silver coinage had become terribly debased by the use of bad and cheap alloys as well as by the prevalent habit of clipping the coins. Things were in a very bad state, even to the extent of English silver being refused by the Bank of Amsterdam—the then financial centre of Europe. Such silver coins to the nominal value of five and a half million pounds sterling were in circulation, the actual value being less than eighty per cent of the nominal value.

In 1694 Charles Montague was appointed Chancellor of the Exchequer, and he decided to restore the silver coinage to its full value. He had conferences with various people, including Newton, on the matter. The House of Lords proposed to cancel all such coins from some given date, so that the sufferers would be

GUARDIAN OF THE COINAGE, 1696–1727

the poorest and those least able to bear it. The House of Commons decided, however, that the nation had to bear the loss, and resolved to issue new and full-valued coins in exchange for the old ones. A vast amount of work thus fell to the Royal Mint.

Montague was an intimate friend of Newton's and President of the Royal Society. He therefore realized that he had in him the right man for carrying through this difficult project. The Wardenship of the Mint in London became vacant early in 1696, and on March 19th of that year Montague wrote to Newton, informing him that the King had appointed him to this post. "The office is the most proper for you. ... 'Tis worth five or six hundred pounds per annum, and has not too much business to require more attendance than you may spare. I desire you will come up as soon as you can. ... I believe you may have a lodging near me." Newton accepted the post, and took over his new functions without delay.

Although the Wardenship of the Mint had become a sinecure and incumbents hardly ever "condescended to come near the Tower," Newton took his new duties very seriously. Hard work was necessary, as the debased silver was causing much suffering and fomenting popular dissatisfaction. "The ability, the industry and the strict uprightness of the great philosopher," says Macaulay, "speedily produced a complete revolution throughout the department which was under his direction." "Well had it been for the publick, had he acted a few years sooner in that situation," said one who had business with Newton and could appreciate his merits at first hand.

For three years Newton laboured in the nation's cause, and in 1699 the recoinage was complete. "Till the great work was completely done, he resisted firmly, and almost angrily, every attempt that was made by men of science, here or on the Continent, to draw him away from his official duties." Instead of producing fifteen thousand pounds' worth of silver per week, Montague and Newton had been able to increase the yield to one hundred and twenty thousand pounds' worth per week. They had to withstand all kinds of opposition, and underhand attempts at bribery and corruption—even open rebellion was threatened in the country. But a gratified nation recognized its benefactors at last. In 1699 Newton was appointed Master of the Mint, a high office worth between twelve hundred and fifteen hundred pounds per annum, which he held for the remainder of his life.

Newton enjoyed his new circumstances for one reason in particular. He was always careless about money. Although scrupulously exact in all business matters, he sometimes suffered great losses by fraud or theft and bore them philosophically. But he enjoyed his wealth because he took a great delight in giving, believing that a man should give during his lifetime and not wait till he is dead. Newton helped his many relatives, both on the paternal side and on the maternal side: he was particuuarly kind to his stepsisters and stepbrother and to their families. Mary Smith got a permanent allowance from him. But it was a daughter of Hannah—the younger sister—who enjoyed Newton's special favour, and played a prominent part in his private life.

Hannah Smith had married Robert Barton of Brigstock, Northamptonshire. In 1679 a daughter was born, called Catherine. She was a very clever girl and showed signs of becoming a great beauty. When Newton went to live in London, Catherine joined him and kept house for him for many years. Newton had no official residence at the Mint, although he apparently lived there in bachelor fashion at first. In 1697 he took a house in Jermyn Street (also written German Street), a street running parallel to Piccadilly,

GUARDIAN OF THE COINAGE, 1696–1727

and connecting Regent Street and St. James Street. He had much entertaining to do, and naturally a beautiful and accomplished young lady, famous for her conversational powers and ready wit, was just the person to help Newton with his social duties.

There is a tendency to regard Newton's transfer to London as a loss to science and as a confession on the part of the British people that the only way to reward scientific genius in one of its citizens is to make him relinquish his scientific work altogether—since no posts exist in the academic sphere that are worth having, and that may be awarded as prizes for merit of the highest order. Much may be said even to-day about the inadequacy of the returns in academic life—in Newton's days they were absurdly inadequate—but it must be pointed out that in Newton's case any academic reward would by its very nature have been unsuitable. He was not exhausted scientifically, but he had made such stupendous scientific discoveries that he could not hope to improve on them. His place was no longer in a college or in a laboratory, but rather in the forefront of his nation's science, putting at the disposal of his people and of humanity his unique gifts as a worker and as a thinker.

From 1665 to 1696 Newton was essentially a scientific researcher, and laboured to unravel the laws of nature. For an equal period, 1696 to 1727, Newton was the servant of his nation and the foremost representative of British learning. No greater mathematician, physicist and astronomer, combined in one person, ever existed, no greater Master and Warden of the Mint ever existed, and no more noble figure could be imagined to act as the presiding genius over British science. Newton's place was in Cambridge for the first half of his adult life, his place was in London for the second half of his adult life, when he had performed his scientific mission to mankind, and had become a figure of national and international renown.

Although Newton desired to devote himself exclusively to the important task that he had in hand at the Mint he could not escape being interested in the scientific problems of the day. He was consulted about the mathematical syllabus at Christ's Hospital, and gave excellent advice. He showed his great mathematical ability by solving in a few hours two problems that the celebrated John Bernoulli put as a challenge to the greatest mathematicians of the world. One of these problems had its origin in optical theory, and deserves mention since it contains the germ of an idea that has played a fundamental part in dynamical theory of the last hundred years. The problem was: Given two points A, B such that the straight line joining them is neither horizontal nor vertical, to find how the curve joining them must be drawn, so that if a particle starts from the top end and falls along the curve under gravity it shall reach the lower end in the least possible time. Newton showed that the curve required is a cycloid, or the path described by a point on the rim of a wheel while the wheel itself rolls along the flat ground. Such a curve is called a brachistochrone. The idea underlying the problem is that of the calculus of variations, and the suggestion that something in nature, say, a length or a time or an energy, is a maximum or minimum has become one of the most fertile methods of dynamical and general physical development.

The work on the motions of the moon lay neglected for a couple of years, but at the end of 1698 Newton resumed it, once more in conjunction with Flamsteed, and once more with considerable unpleasantness in their personal relationships. In 1699, after declining an offer of a considerable pension from the French king—Newton not only disliked gifts to himself, but he was also a very loyal Whig, "the glory of the Whig party," as Macaulay calls him—he was elected a foreign associate of the French Academy, together with the most celebrated scientists of the day—Leibniz, James and John Bernoülli, and Roemer.

In 1699 Newton decided that he could not continue as Lucasian professor, and appointed a deputy in that remarkable mathematician and theologian, William Whiston, the translator of Josephus. Whiston was very unorthodox in religious belief, but with Newton's help he was elected Lucasian professor, when in 1701 Newton finally resigned his chair and his Trinity fellowship. As a matter of fact, Whiston was expelled from Cambridge in 1710 for his heterodox views.

Newton still had a bond of union with Cambridge. In November, 1701, he was once more elected M.P. for Cambridge University, and sat till the dissolution in July, 1702, after the death of King William III, and the succession to the throne of Queen Anne. Newton did not seek re-election for some years.

In scientific matters Newton's activity was uninterrupted. In 1699 he drew up a scheme for a reformed calendar, with quite revolutionary suggestions, based upon an ingenious attempt to correlate the calendar with the actual equinoxes and solstices. In 1700 Newton sent to Halley a description of an instrument he had invented for measuring the angle subtended at the eye by two inaccessible points. The instrument was, in fact, the sextant, so important in navigation and surveying: but nobody realized its value, and John Hadley reinvented it in 1730—several years after Newton's death. In 1701 Newton published a paper on scales of temperature, incorporating two discoveries of great importance. The first was his law of cooling, which is that the rate of cooling at any moment of a warm body is proportional to the difference between the temperature of the body and of the surrounding medium, say, air. The second was that while a body is melting or evaporating its temperature remains constant. Both discoveries play important rôles in modern physics.

Newton's standing in the world of science was unchallenged. The great Leibniz said of him in 1701 that his work in mathematics was worth more than all that had been done before him. This was an exaggeration, perhaps, and Leibniz' relations to Newton later on did not suggest such a pitch of admiration. In England there was no doubt about Newton's scientific stature. The Royal Society was the established and recognized repository of all that was best in English science, and the post of President carried with it the status of pre-eminence in the world of knowledge. In November, 1703, Newton was elected President of the Royal Society. He was at last in the position, as it were, preordained for him. Most gifted of all Englishmen, supreme in knowledge and discovery, standing at the summit of happy achievement, and enjoying universal approbation and renown, Isaac Newton had risen from the obscurity of a county hamlet to be the glory of his nation and a blessing to mankind.

CHAPTER XI

THE DOYEN OF BRITISH SCIENCE, 1703-1727

Honour is not repugnant to reason, but may arise therefrom.
Baruch Spinoza : " Ethics "

NEWTON had become a prominent national figure. He moved in the best society and lived in considerable style. He employed many servants, kept a carriage, and entertained hospitably and well. His salary was more than he needed and he was able to exercise his charitable propensities to the full. He was ever ready to support the scholar, and he showed special favour to men in whom he discerned mathematical ability of a high order.

The friendship that Newton enjoyed with the House of Orange was cemented in various ways. When King William III died in 1702 he was succeeded by his sister-in-law, Anne, the daughter of ex-King James II. Her husband was the good-natured Prince George of Denmark, who was elected to the Royal Society. As President, Newton came into contact with the Prince Consort and encouraged him in his design to do something tangible for science. Prince George offered to pay the cost of printing Flamsteed's observations, including his star catalogue, and asked Newton, together with Wren and several other fellows of the society, to act as referees in the matter, and to decide which observations were to be printed and the order in which they were to be arranged. By the beginning of 1705 the referees had reported and the work of printing could begin.

Queen Anne became interested in these matters, too, and learnt to see in Newton the man of highest genius in her land. On April 16, 1705, she visited Cambridge and held a court at Trinity Lodge. She took advantage of the occasion to confer upon Newton the dignity of knighthood.

Yet royal favour carried disappointment in its train. Parliament had just been dissolved. In the new elections Newton was prevailed upon by his Cambridge friends to offer himself for election once more. But the Tories were in the ascendant, and Newton suffered ignominious defeat only a month after he had been dubbed a knight. With four candidates in the field he was returned bottom of the poll. Newton never sought parliamentary honours again—no real loss to the nation, since he was not cut out for parliamentary life or for a political career. He had always been shy of public discussion, even in matters scientific, and he never bore with a good grace criticism and opposition.

Of Newton's private life we do not know much. He seems to have had close friends in other than royal circles, too, and he is credited with having written a very oblique sort of proposal of marriage to a Lady Norris—a much-married lady who had survived three husbands. If the proposal was a reality it had no serious sequel. Newton was then sixty-three years of age, and he remained a bachelor to the end of his days. His niece Catherine continued to keep house for him till she married.

The most important events in the remainder of Newton's life were of a scientific nature. Two important works by him were published soon after his elevation to the Presidency of the Royal Society. The first was on "Optics," and contained a connected account of all Newton's labours in this subject. The manuscript was ready for several years, but Newton did not wish to have it printed while Hooke lived, as he did not desire to have any further unpleasantness with him. Hooke died in 1703 and Newton's "Optics" was published in 1704.

Unlike the "Principia," the "Optics" was published in English: a Latin edition appeared in 1706. The book was a great success: it enjoyed widespread approval, and several editions both in English and in Latin were printed within a comparatively few years. As is to be expected, the book deals with the composite nature of white light and with the cause of colour in bodies, with reflection and refraction upon the basis of Newton's corpuscular view of light, with colours in thin films and the theory of fits of preference for reflection and for refraction, with diffraction (called inflection), with the rainbow, with halos, etc. But in the first edition of the "Optics" two mathematical treatises, quite irrelevant to the main purpose of the book, were added. One was on "Quadratures of Curves," or really the theory of fluxions as used in finding areas enclosed by curves: the second was an account of seventy-two curves of the third order, so-called cubic curves. If the inclusion of the first paper was expected to make clear Newton's position as the inventor of fluxions, then the effect produced was the very reverse. In the review of the "Optics" in Leibniz' organ, the "Acta Eruditorum," written by Leibniz himself, but not signed, it was quite broadly hinted that Newton had simply taken Leibniz' published work on the calculus, and rewritten it in the fluxionary notation!

Newton lectured on algebra while Lucasian professor, but he had not published his notes on the subject—he was never anxious to display his pure mathematics, which he looked upon as a tool for physics and astronomy. William Whiston had attended these lectures in 1685 and so possessed them in manuscript : he had them printed in 1707. The book is in Latin and is called " Universal Arithmetic." This book, too, was received with great approval both in England and on the Continent. An English translation appeared soon, while a second edition was issued within five years of the first.

Meanwhile, the printing of Flamsteed's observations went on under Newton's supervision. A first volume was published in 1707, but the death of the Prince Consort in 1708, and Flamsteed's own dilatoriness, caused a delay of several years in the publication of the remainder. Newton and Flamsteed fell out over this again and again. The way in which Flamsteed carried on his work at the Royal Observatory was not considered satisfactory, and finally the Royal Society asked the Queen to place the Observatory under some responsible supervision. At the end of 1710 such a Board was constituted, with Newton as its chief member, to examine the instruments at the Observatory and to put them into proper repair. A violent quarrel broke out between Newton and Flamsteed—the latter resenting any interference with his job, in which he had been working for over thirty years, with only one hundred pounds per annum to cover salary, instruments, running expenses, etc. Newton lost his temper, and Flamsteed's subsequent behaviour was almost vandalistic. The scientific contact between Newton and Flamsteed did not survive this shock ; they hardly corresponded again till Flamsteed's death in 1719. In any case, Newton had obtained sufficient information about the moon for his purposes—namely, a second edition of the " Principia."

This work had become so scarce that Newton had more than once been asked by Bentley to allow him to reissue it. Newton had made a number of notes in the margin of a copy of the first edition, and in

June, 1708, he sent this to Bentley, who at once proceeded with having the work reprinted. But Bentley became involved in quarrels at Trinity and Newton was too busy to see to the printing himself. In the following year a brilliant young mathematician, Roger Cotes, a fellow of Trinity College, who had been appointed the first occupant of the Plumian Chair in astronomy and experimental philosophy at the early age of twenty-five, was entrusted with the task of preparing the second edition of the "Principia." Newton still had some corrections and additions to make and in October, 1709, Cotes received Newton's corrected copy through William Whiston. Newton was busy moving into a new residence in Chelsea when he had the satisfaction of hearing from Bentley: "You need not be so shy of giving Mr. Cotes too much trouble. He has more esteem for you, and obligations to you, than to think the trouble too grievous; but, however, he does it at my orders, to whom he owes more than that, and so pray you be easy as to that. We will take care that no little slip in a calculation shall pass this fine edition. . . ."

The choice of Cotes was a happy one. He set himself to the task with energy and with insight. He was not content to "print by the copy sent him," as Newton had suggested, but to produce a properly edited work. He consulted with Newton over and over again as the work of printing went on, and the long correspondence between Newton and Cotes constitutes one of the most valuable sets of scientific letters extant. Several additions were incorporated in the new edition of the "Principia"; as Newton wrote in his preface: "In the Third Book the theory of the moon, and the precession of the equinoxes, are more fully deduced from their principles, and the theory of comets is confirmed by several examples, and their orbits more accurately computed."

The production of the second edition of the "Principia" occupied four years (1709-1713). In 1710 Newton changed his residence again, moving to St. Martin's Street, which runs into Leicester Square. The house he took was then the first one on the left as one enters the street from the square: it was of some size and in a suitable position for one who was continually receiving visits from distinguished Englishmen and foreigners. His niece, Catherine Barton, had by now attracted the attention and admiration of numerous famous men like Dean Swift and Charles Montague—the latter had become Earl of Halifax, and is said to have wanted to marry her; but Miss Barton remained unmarried for some years yet and still served Newton as the leading figure in his household. Newton himself was already an old man, nearly seventy years of age. He was never above middle size, and tended to corpulency as he grew older. He was in good health, but somewhat clumsy in his attitudes. He was always friendly, but being continually steeped in meditation he spoke but little in company. He was by no means eccentric, but he was as apt to be remiss in his social duties as negligent of his own comforts. That he sometimes sat in bed forgetting to dress himself through absorption in some problem merely proves him as human as all of us, not as human, indeed, as Descartes who, at one period of his life, spent most of his time in bed. If Newton sometimes thought he had fed when some kind friend had eaten his lunch for him, it only proves that he was no slave to food, just as he had remained free of the serfdom to snuff and tobacco.

In 1711 Newton published an important paper on mathematical computation, the "Methodus Differentialis," containing formulæ that have not lost their importance to this day. But a storm was brewing that was soon to break over Newton's head with great violence. His negligence with respect to his discovery of fluxions was bringing its punishment in its train.

As early as 1699 a Swiss mathematician, Fatio de Duillier, had somewhat clumsily broken a lance in defence of Newton's priority in the discovery of fluxions. " Compelled by the evidence of facts," he said, " I hold Newton to have been the first inventor of the calculus, and the earliest by several years. And whether Leibniz, its second inventor, has borrowed anything from him, I would prefer to my own judgment that of those who have seen the letters of Newton and copies of his other manuscripts." When Newton's "Optics" appeared and the reviewer in the "Acta Eruditorum."—who was, in fact, Leibniz himself, under the cloak of anonymity—implied that Newton had plagiarized Leibniz, the fury of English mathematicians was roused. John Keill of Oxford replied in 1708, and insinuated quite plainly that Leibniz had obtained his differential calculus from some of Newton's manuscripts to which he had been allowed access. When Leibniz read this he sent a strong protest to the Royal Society and asked that body to make Keill disown publicly the implied accusation. Keill did not disown his remarks, but drew attention to the review in the " Acta Eruditorum," and explained that from what Leibniz saw of Newton's letters to Oldenburg he could discover the idea of the calculus. Leibniz was not satisfied, and in December, 1711, he stated openly that the views expressed by the reviewer of the " Optics " had been only fair and just.

It was now open war : not only was Keill involved but the Royal Society and its President, Newton, were thus dragged into the dispute. It was a very unpleasant affair as all disputes involving personalities must be. Attempts have been made to give deep motives to the disputants and recondite causes to the bitterness of feeling evinced. One theory is that

Leibniz and Newton were really engaged in a political quarrel—that Newton was a fanatical Tory, and as such opposed to the Elector of Hanover, so that Leibniz naturally also shared in this displeasure. This pretty theory is, however, inconsistent with Newton's being "the glory of the Whig party," in the words of Macaulay, and extraordinarily friendly with the Elector of Hanover when he became King of England only a few years later, especially with the Princess of Wales, who was a particular friend of Leibniz. No: it is not necessary to invent such impossible theories. Newton and Leibniz were great men but only men, subject to the passions and failings of the species.

The Royal Society had to take some action, and in March, 1712, a committee was appointed to examine all available documents and to express an opinion as to the independence and priority of the discovery of the calculus idea and method by Newton and Leibniz. It is a great pity that Newton was himself President at the time—an obvious charge of partiality and prejudice lay at the door of the committee and would discount much of the value of its decision. Among the members were Halley and the mathematicians Machin, de Moivre and Brook Taylor—great names indeed—as well as the Prussian Minister and others. There can be little doubt that the committee tried to be as impartial as possible: at the same time it is clear that it was not an independent jury, but rather a body appointed to vindicate Newton.

Halley had been many things since Newton had come to London. Newton helped him to a post in one of the branch mints. Halley then undertook various scientific voyages, and finally became Savilian professor of geometry at Oxford. It was in his handwriting that the report of the committee was presented. After giving a brief history of the various papers and letters written by Newton the committee reports: "That the differential method is one and the same with the method of fluxions, excepting the name and mode of notation. ... And therefore we take the proper question to be, not who invented this or that method, but who was the first inventor of the method. ... For which reasons we reckon Mr. Newton the first inventor; and are of opinion that Mr. Keill, in asserting the same, has been noways injurious to Mr. Leibniz. And we submit to the judgment of the Society, whether the extracts, and letters, and papers, now presented, together with what is extant to the same purpose, in Dr. Wallis' third volume, may not deserve to be made public."

It must be said in all fairness that the committee's report was a very balanced and reasonable one. Its suggestion to print the "extracts, letters and papers" was adopted, and in 1713 there appeared the "Commercium Epistolicum" or Correspondence with John Collins and others concerning the development of the method of the calculus.

When Leibniz heard of this publication he became furious, and a very acrimonious correspondence ensued, in which Leibniz as well as the great John Bernoulli figured very unsatisfactorily, while Keill, acting no doubt under Newton's directions, laboured valiantly in defence and offence. Meanwhile, the second edition of the "Principia" appeared in 1713, and a second edition of the "Optics" in 1714. An attempt was made by Tory intriguers to remove Newton from the Mint by the offer of a very liberal pension: Newton declined, and remained in active charge of this institution. His opinions were sought after on matters of moment to the country, and when Queen Anne died in 1714 and was succeeded by the Elector of Hanover as King George I—the "honest blockhead" who punished so brutally in his cousin-wife the sin that he himself practised so flagrantly—Newton remained a favourite at court. The King's son, George, was declared Prince of Wales; the Princess of Wales was the accomplished Caroline of Anspach, who had regular evenings when distinguished men of science and learning visited her salon. Needless to say, Newton was an especially welcome visitor on such and other occasions.

The weary quarrel over the calculus dragged on. The new editions of the "Principia" and the "Optics," and the succession of the Hanoverian house to the English throne, suggested to Leibniz a new weapon of attack. One of the most important changes in the second edition of the "Principia" was a general scholium at the very end of the work. Here Newton points out very forcibly the inadequacy of the vortex theory of Descartes, and then proceeds to a semi-scientific, semi-theological disquisition on the nature of God, finishing with a slight reference to his ether theory of gravitation. To the new "Optics," too, Newton added a number of "queries" containing views and speculations on the constitution of matter in its various manifestations, on gravitation, etc. Leibniz pounced upon these things as a means of discrediting his enemy. He had already, in 1710, attacked Newton's theory of gravitation as introducing the occult and the miraculous into natural philosophy. He renewed this attack towards the end of 1715, and directed his arguments to the Princess of Wales. He declared that philosophy and natural religion had decayed among the English, and accused Newton of materializing the conception of God and limiting His omnipotence.

At the instance of the King, Newton now intervened personally in the dispute. This led Leibnitz to yet another attempt to overcome Newton. He challenged Newton in 1716 to solve a problem which is only

DOYEN OF BRITISH SCIENCE, 1703-1727 147

soluble by a person in complete possession of the ideas and methods of the calculus. Newton solved the problem one evening after returning from his duties at the Mint. The dispute would perhaps have dragged on indefinitely, but Leibniz" death in November, 1716, put a stop to the quarrel—in the sense at least that Leibniz was silenced and the Englishmen had it all their own way.

Gottfried Wilhelm Leibniz was three or four years younger than Newton. He was a man of remarkable genius as an administrator, diplomat, philosopher and mathematician. He was the outstanding German savant of his day, a man who has left his mark on metaphysical inquiry and in pure mathematics. His mathematical genius was for analytical form rather than for the geometrical and physical intuition in which Newton excelled. To him the calculus was more of a mathematical exercise with a beautiful notation—to Newton it was a rugged tool, practical and penetrating. Newton realized the ideas better, Leibniz devised the polished processes.

This makes the dispute all the more regrettable. The effect of the dispute was to produce an estrangement between the British mathematicians and their continental colleagues. Newton's successors in England cried to follow his geometrical methods. They failed to make any real progress, and the Cambridge school of mathematics declined and practically died with Newton. Cambridge became the home of mathematical teaching with an examination tradition that evolved into the Mathematical Tripos : but it was not till a century after the end of the Newton- Leibniz controversy that Cambridge mathematicians began to show signs of originality and fertility—and this happened only when the prejudice against the continental Leibnizian notation and methods was overcome and the slavish imitation of Newton abandoned.

Newton and Leibniz were independent thinkers
and discoverers. It was Newton's own fault and the
fault of Barrow, Collins and other friends of his youth,
that he was embroiled in controversy over the calculus.
It is a recognized principle that priority is decided by
the date of publication, and not by the date of discovery in the privacy of one's study. This was
acknowledged in Newton's time, too. Nobody denied
to Hadley the merit of and credit for the sextant,
although Newton had invented it thirty years earlier:
Hadley made the discovery public and reaped the
reward. Why should not the prior publication by
Leibniz have been treated in the same manner?
Newton had too little respect for his own mathematical
powers, and lived to regret it.

During the progress of the calculus controversy
Newton suffered two severe losses of friends whom he
esteemed highly, Charles Montague, Earl of Halifax,
and Roger Cotes. The former had reached the highest
positions in the State and had done much to re-estab-
lish English finance. He started the National Debt
and the Bank of England, and reformed the currency
with Newton's help. He had been first minister of
the Crown under King William III, and although
forced to retire from much of his public life when the
Tories were the dominant power in the land, he emerged
again as first minister of the Crown under George I.
His death occured in May, 1715. He had been
a widower for some time—some say that he had
married Newton's niece—and he bequeathed one
hundred pounds to Newton as a mark of honour and
esteem, and five thousand pounds with various other
benefits to Catherine Barton " as a token of sincere
love, affection, and esteem I have long had for her
person, and as a small recompense for the pleasure and
happiness I have had in her conversation."

Halifax was only fifty-four years of age when he

died. Even severer and more tragic was the death of Roger Cotes in 1716, who was taken away when not yet thirty-four years old. "If Mr. Cotes had lived," Newton said, "we might have known something."

Two years after the death of her admirer, Halifax—whether this is significant or not, one cannot say—Catherine Barton, then thirty-eight years of age, married John Conduitt, a friend of Newton's, nine years her junior, who enjoyed independent means and was a member of Parliament. A year later Newton had the joy of welcoming into the world a grand-niece, Catherine Conduitt, who later became Viscountess Lymington and ancestress of the family of the Earl of Portsmouth.

A semi-domestic event which must have given Newton much pleasure was the elevation of Halley to the position of Astronomer Royal in 1720: no doubt Newton had a share in making this appointment. Halley was Secretary of the Royal Society since 1713, so that Newton had a great source of satisfaction in the scientific career of this sincere friend and admirer.

Newton was now nearly eighty years of age. He was not frail or obviously aged. We have a portrait of him at the age of eighty-three, and even then he looked well—a venerable sight with fine silver-white hair. He never became bald, never wore glasses, and enjoyed the use of his own teeth to the end. His mind was lively and alert, and he carried out his official duties at the Mint and his functions as President of the Royal Society without interruption. But in 1722 Newton began to suffer from the infirmities of old age. It was an affection of the bladder, which turned out to be much less serious than at first appeared. But Newton had to take special precautions with his food, to give up exciting dinner parties, and to give up the use of his carriage, since jerking motion was liable to produce another attack. In this way he remained tolerably well for about two years.

Newton used this respite for a very important purpose. A third edition of the "Principia" was needed. It is remarkable that for each edition of this great work Newton enjoyed the inestimable advantage of the printing being carried out under the direct supervision of a young and energetic man of learning and talent. Halley was only thirty years of age when he brought out the first edition in 1687; Cotes was barely thirty-one years of age when he brought out the second edition in 1713. For the third edition Newton secured the help of another brilliant young scientist, Henry Pemberton, who began the task in 1722 when he was twenty-nine years old, and published the third edition in 1726 when he was barely thirty-two.

Henry Pemberton was a young doctor who had become acquainted with the "Principia" when still a medical student. This made him turn his attention to mathematical studies, and he became acquainted with Newton, who recognized in some of Pemberton's work considerable mathematical power. The old scientist and the young doctor became great friends and finally the latter undertook to issue the new edition of the "Principia."

Pemberton, like Cotes, was not an editor in the restricted sense. He discussed with Newton every point that required consideration during the course of the printing, and did everything with the knowledge and approval of his senior. Meanwhile more obvious signs of physical decay made themselves manifest in Newton. In August, 1724, he suffered from stone in the bladder and in January, 1725, he was attacked by inflammation of the lungs, accompanied by severe coughing, followed a month later by gout in both feet. The air of the Leicester Square district was not considered good for him, and Newton moved to Orbell's Buildings, now called Pitt's Buildings, in Pitt Street, near High Street, Kensington. The change did him some good, but he felt that he ought to give up his duties at the Mint. He wanted, apparently, to keep the Mastership in the family, as he tried to get Conduitt appointed Master. But this did not materialize, and Newton remained Master—but Conduitt did most of his work there! Newton continued to preside at the Royal Society, but it is on record that he fell asleep while occupying the chair.

Newton always befriended mathematical genius—he encouraged and helped men like Cotes, Duillier, David Gregory and Pemberton. In 1725 two Scotch mathematicians received marks of his approval and his kindly help. Colin Maclaurin was one of the greatest mathematicians that Scotland has produced. In 1717, at the early age of nineteen, he was appointed professor of mathematics at the Marischal College, Aberdeen. But Maclaurin desired to go to Edinburgh, and to obtain the reversion of the chair held by James Gregory —the nephew of the inventor of the reflecting telescope. It was Newton's recommendation that in 1725 secured for Maclaurin the post of assistant and successor to Gregory, and he even offered to contribute twenty pounds per annum to the salary of the assistantship in order to make the arrangement feasible. Maclaurin became professor a year later.

James Stirling was a brilliant young Oxford student of mathematics. In 1715, when twenty-three years of age, he got embroiled with Jacobites and had to flee. He went to Venice and soon began to publish important work on geometry. Newton not only got Stirling's work published in England, but in 1725 obtained permission for him to return to England. Stirling had an interesting career in Scotland, finishing off as the manager of a coal-mine. Newton's approval of

Maclaurin and Stirling has been more than endorsed by posterity.

The third edition of the " Principia " was meanwhile going ahead, and early in 1726 it appeared, containing an excellent portrait of Newton at the age of eighty-three. Newton's generosity to Pemberton was both moral and material. He presented him with a sum of two hundred guineas (he had already given a hundred guineas to the astronomer Pound, whose observations Newton used in the new edition) and paid him a high compliment in the preface. Nobody ever could complain of Newton as regards personal generosity. He even remembered to contribute towards the cost of repairing the floor of Colsterworth Church.

The issue of the third edition of the " Principia " was the end of Newton's scientific work.

CHAPTER XII

THE END, 1727

I do not know what I may appear to the world; but to myself I seem to have been only like a boy playing on the sea-shore, and diverting myself in now and then finding a smoother pebble or a prettier shell than ordinary, whilst the great ocean of truth lay all undiscovered before me.

ISAAC NEWTON

A REMARKABLY useful and well-spent life was drawing to a close. The great worker and thinker felt that the end was near. His calm objectiveness in scientific judgment never deserted him. Conscious of the unprecedented progress that he had made in the understanding of Nature's laws, he was, nevertheless, ever ready to revise his views when facts seemed to go against them. When, a few months before his death, he was threatened with the complete demolition of his structure based on universal gravitation, his reply was simply: "It may be so, there is no arguing against facts and experiments." Two centuries have passed and the structure stands firmer than ever.

The last years of Newton's life were occupied with another interest. The study of chronology is a natural occupation for a mathematician, and several great mathematicians have worked at it: the different races and religions of humanity have seen to it that sufficient confusion exists to puzzle the most ingenious. Newton had often read history and chronology as a mental recreation, and he was induced by the Princess of Wales to write a brief account of his own views of the correlation between the dates of events in ancient history. When Newton found in 1725 that this had got into print without his permission, he decided to prepare a large work on the subject. The " Chronology of Ancient Kingdoms " was completed only a short time before his death and was not published till a year after his death.

In August, 1726, further physical ailments attacked Newton. He was confined to his house at Kensington, and spent much of his time in reading, especially the Bible. On Tuesday, February 28, 1727, he felt, as he thought, well enough to go to Crane Court, near Fleet Street, London to preside at the Royal Society on Thursday, March 2nd. He went, and came back to Kensington on Saturday, March 4th, seriously ill. This was the beginning of the end. A week later the illness was diagnosed as stone in the bladder, and it was felt that death was near. Newton suffered excruciating pains, but he bore them patiently and uncomplainingly. In the intervals he conversed cheerfully, and on Wednesday, March 15th, there appeared to be some improvement in his condition. This was an illusion, however. On Saturday evening, March 18th, he had a relapse and became insensible. He remained like this till after midnight of the following day, and died painlessly between one o'clock and two o'clock in the morning of Monday, March 20, 1727.

Newton died in his eighty-fifth year. The highest possible posthumous honours were done him. His body lay in state in the Jerusalem Chamber on Tuesday, March 28th, and was buried in Westminster Abbey. A plain flagstone marks the actual grave with the inscription " Hic depositum est Quod Mortale fuit Isaaci Newtoni." A monument dedicated to Newton's memory was erected in 1731, when also a medal was issued at the Mint in his honour.

Newton left his family estates at Woolsthorpe and Sewstern to the heir-at-law, the ne'er-do-well, John Newton, his second cousin. He had already in his

THE END, 1727

lifetime made ample provision for more distant relatives like the Ayscoughs, and for such nearer relations as were beyond the scope of his bequests by will. His personal estate of thirty-two thousand pounds (parlty invested in South Sea Stock) went to his eight living nephews and nieces, the children of his stepbrother, Benjamin Smith, and of his stepsisters, Mary Pilkington, and Hannah Barton. The Mastership of the Mint went to John Conduitt after all.

The impression that Newton made on British scientific life, and on the scientific life of humanity as a whole, has remained one of the indelible imprints on the sands of time. It was only after his death, as generation after generation of the mightiest intellects of Europe followed out his ideas, and triumph after triumph was scored in the elucidation of Nature's puzzles, that the true greatness of his genius was appreciated. Rarely, if ever, have so many noble qualities been united in one man. A great experimentalist and manipulator, a profound mathematician and theorist, a clear and logical thinker, a fine exponent and writer, a practical national benefactor, a kind and appreciative supporter of merit and ability, a model relation to the members of his family, an example of unerring rectitude, an abhorrer of all vice and of cruelty to man and beast—Newton has long since obtained absolution at the hands of posterity for the faults of temper which sometimes marred his personal relationships, and the faults of self-appreciation that ignored the mathematician in himself and sought inspiration in the pseudo-science of alchemy. Humanity will ever rejoice to have counted Newton among its sons.

SIBI GRATULENTUR MORTALES, TALE TANTUMQUE
EXTITISSE
HUMANI GENERIS DECUS.

NOTE

Many editions of Newton's chief individual works, such as the "Principia" and "Optics" have appeared from time to time, and are easily obtainable. No really authoritative edition of his complete works has been issued yet. A collection of Newton's writings in five volumes was issued by Samuel Horsley in the years 1779–1785, under the title: *Isaaci Newtoni Opera quae extant Omnia.*

INDEX.

Aberration, chromatic, 59, 60
  spherical, 54–7
Academic life, 133
Acceleration, 44
Achromatism, 68
"Acta Eruditorum," 108, 139, 143
Action at a distance, 122
Alchemy, 125, 155
Algebra, 65, 139
Anne, Queen, 135, 137–8, 145
Apocalypse of St. John, 125
Apple, 45, 46
Apples for cider, 82
Arabs, 32
Areas, 21–3
Aristotle, 43
Astrophysics, 73
Attraction of sphere, 110
Authority, 43, 69, 71, 79
Ayscough, 5, 7, 155

Bacon, Roger, 3
Barrow, Isaac, 58, 64–5, 148
Barton, Catherine, 9, 132, 142, 148–9
Bentley, Richard, 125–6, 140–1
Bernoulli, James, 135
Bernoulli, John, 134–5, 145
Binomial Theorem, 16–8, 26
Bohr, Niels, 73
Boyle, Robert, 26, 74, 78, 101, 125
Brachistochrone, 134
Bradley, James, 77
Brahe, Tycho, 11, 38–9, 43, 110
Brigstock, 132
Burton Coggles, 7
Byron, Lord, xii

Calendar, 4–5, 135

Cambridge, 16, 123, 138, 147
Caroline of Anspach, Princess of Wales, 144, 146, 153
Cartesian ovals, 56, 60,
Cassini, Giovanni Domenico, 114
Cavalieri, Bonaventura, 25
Central acceleration, 48–9, 83, 108
Centre of gravity, 106–7, 114
Charles I, 1, 2
Charles II, 1, 65–6, 71, 90, 92–3
Christ's Hospital, 134
Chronology, 153–4
Clark, apothecary, 7
Coinage, 130
Collins, John, 64–5, 145, 148
Colour, 29, 53, 57–8, 67, 78, 139
Colsterworth, 2, 5, 152
Comets, 9, 88, 90, 93, 117, 119–21
Commercium Epistolicum, 145
Conduitt, John, 149, 151, 155
Conduitt, Catherine, Viscountess Lymington, 149
Conics, 38–9, 108, 117
Cooling, law of, 135
Copernican system, 37, 40–1
Copernicus, Nicolas, 10, 35–8, 41, 104
Corpuscular theory of light, 77, 139
Cotes, Roger, 141, 149–51
Cromwell, Oliver, 2, 3, 10
Cubic Curves, 139
Cycloid, 134

De Duillier, Fatio, 143, 151
Deferent, 32, 35–6
De Moivre, Abraham, 144
De Motu, 91, 94, 107
De Mundi, 96, 107, 112, 117

De Revolutionibus Orbium Coelestium, 38
Descartes, René, 16, 42, 56–8, 122, 142, 146
Diamond, 127
Differential Calculus, 82, 108, 139, 143
Diffraction, 78, 139
Discs of planets, 37
Dispersion, 59, 68–9
Dispute with Leibnitz, 143–7

Earthly and celestial, 33, 38, 93, 113
Earth, motions of, 35–6, 85, 87
Eclipse, 118
Edinburgh, 123, 151
Einstein, Albert, 73, 119
Elasticity, 45
Electricity, 81
Ellipse, 39, 40, 83–4, 90, 117, 119
England, 122, 136
Epicycle, 32, 35–6, 38
Ether, 42, 77–9, 122, 146
Exhaustions, method of, 23

Falling body, 44, 86
Fermat, Pierre de, 25
Fire, 127
Fits, 76–8, 139
Fixed stars, 106
Flamsteed, John, 88, 93, 101, 114, 128–9, 134, 137, 140
Flattening of earth and planets, 94, 111, 114
Fluids, 111
Fluxions, 19, 21–2, 27, 43, 64, 70, 81–2, 107–8, 125, 139, 143
Foucault, Jean Bernard Léon, 77
French Academy, 135
Fresnel, Augustin Jean, 79

Gale, experiments in, 10
Galilei, Galileo, 37, 42–4, 46, 53, 56–8, 105
George, Prince of Denmark, Prince Consort, 137, 140

George I, Elector of Hanover, 144–5
George, Prince of Wales, 146
Grantham, 2, 7, 10, 11, 99
Gravitation, 42, 73, 78, 81, 83–4, 93, 113, 119, 122, 125, 146
Greeks, 33, 38
Gregory, David, 151
Gregory, James, 60, 63
Gregory, James (the younger), 151
Grimaldi, 78
Gyroscope, 114

Hadley, John, 135, 148
Halos, 53, 139
Halley, Edmund, 89–91, 94–6, 101, 119–21, 123–4, 128–9, 144, 149, 150
Hero worship, 79
Herschel, William, 126
Homer, 44
Hooke, Robert, 45, 69, 74, 77–9, 82–4, 86–90, 95–6, 99, 139
Huygens, Christian, 69, 77, 79, 81, 99
"Hypotheses," 113

Interference, 77
Inverse square law, 45, 47, 49, 50, 83–4, 87–8, 90–2, 96, 114

James II, 97, 101, 123, 137
Jeffreys, Judge, 98, 121
Jews, 32

Kant, Immanuel, 126
Keill, John, 143, 145
Kepler, John, 11, 25, 38–42, 48, 57, 104, 110
Kepler's Laws, 40–1, 48–9, 84, 87, 90, 113
King's College, Cambridge, 124
King's School, Grantham, 2, 7, 10–11

Lagrange, Joseph Louis, 118

INDEX 159

Laplace, Pierre-Simon, 102, 118, 126
Latent heat, 136
Laws of motion, 42–3, 45, 92, 103–4, 109
Leibniz, Gottfried Wilhelm, 82, 99, 108, 122, 125, 135–6, 139, 143–8
Librations of moon, 81, 114
Locke, John, 124, 127, 129
London, 16
Lucas, 68
Lucretius, Titus Carus, 126

Macaulay, Lord, 123, 131, 135, 144
Machin, John, 144
Maclaurin, Colin, 151–2
Market Overton, 5
Mars, 38–9
Mary II, 123
Masses, 118
Mathematical Tripos, 147
Mercator, Nicolas, 81
Mercury, 48, 119
"Methodus Differentialis," 142
Milton, John, 102
Mint, Royal, 16, 131–3, 151, 155
Monochromatic light, 74
Montague, Charles, Earl of Halifax, 124, 129–32, 142, 148–9
Moon, 30–1, 37, 88, 101, 113, 115–6, 127–8, 134
Moon's distance, 36, 50
Moon's motion, 46, 50–1, 89, 93

Nature, benevolence of, 105, 107, 109
Nebular hypothesis, 126
Neptune, 119
Neville's Court, Trinity, 112
Newton, Humphrey, 99
Newton, Isaac :
    birth, 2, 4
    family, 5, 9, 155
    baptism, 5
    grandmother, 7

Newton, Isaac (contd.) : .
    school, 8
    games, 9
    farmer, 10
    back at school, Miss Storey, 11
    Trinity College, Cambridge, 12, 15, 19, 62
    scientific character, 15, 16, 47, 51, 77, 92, 99, 129, 155
    B.A., 19
    plague, 21, 29, 53, 62
    falling apple, 45
    optics, 60, 62, 66–7, 74–6, 78
    gravitation, 60, 81
    Fellow of Trinity, 62–3
    M.A., 63
    reflecting telescope, 63, 65, 67
    Lucasian professor, 64–5
    fluxions, 64
    Fellow of Royal Society, 66
    neglect of food and sleep, 70, 99, 126
    excused holy orders, 71, 92, 100
    excused fees by Royal Society, 71–2
    Wave-theory of light ; theory of fits, 76–7
    Hooke, 82
    Halley, 90–1, 94
    writes "De Motu," 91
    "Principia," 94, 101, 122–3, 125, 140–2, 145–6, 150, 152
    writes "De Mundi," 96
    bachelor life, 98–9, 124, 126
    lectures, 99
    garden, 100
    Leibnitz, 82, 122, 143–7
    Parliament, 123, 135, 138
    Charles Montague, 124, 129–32, 142, 148
    mother's death, 124
    lack of recognition, 124
    theology, 125
    unwell, 126–7
    story of dog Diamond, 127
    Locke, 129
    Warden of Mint, 131
    Master of Mint, 132
    money matters, 132, 155
    residences, 132, 142, 151

Newton, Isaac (*contd.*):
  niece, Catherine Barton, 9, 132
  solves problems, 134, 147
  offered pension by French king, 134
  elected to French Academy, 135
  resigns fellowship and chair, 135
  President of Royal Society, 136
  life in London, 137
  knighted by Queen Anne, 138
  proposes marriage, 138
  Roger Cotes, 141
  refuses to retire from Mint, 145
  mathematics, 147, 155
  John Conduitt, 149, 151, 155
  illnesses, 149, 150
  Henry Pemberton, 150-2
  on himself, 153
  last illness, death, and burial, 154
Newton, John, heir-at-law, 7, 154
Newton's rings, 74-6
Norris, Lady, 138
North Witham, 5

Oldenburg, Henry, 66, 68, 82, 96, 143
"Optics," 79, 138-9, 143, 145-6
Orthodoxy, Christian, 126
Oscillatory motion, 110
Oxford, 16, 123, 143-4, 151

Paget, 101
Parabola, 117
Parliament, 1, 3, 123, 135, 138
Pemberton, Henry, 150-2
Pendulum, 111, 114
Pepys, Samuel, 96, 124
Periodicity in light, 76-7
Perturbations, 109, 110, 115
Phases of Venus, 37
Picard, Jean, 84, 89
Planets, 31

Plague, Great, 21, 29, 53, 62
Plato, 126
Portsmouth family, 149
Pound, 152
Precession, 93, 111, 114
Principia, 94, 101, 122-3, 125, 140-2, 145-6, 150, 152
Priority, 148
Prism, 19, 53, 57-9
Prophecies of Daniel, 125
Ptolemaic system, 31-2

Queries in "Optics," 146

Radius of earth, 50, 52, 84, 89
Rates of change, 22
Reflecting telescope, 53, 60, 63, 65, 67, 83
  microscope, 67
Refraction, astronomical, 128
Relative motion, 36, 105
Relativity, 73, 119
Resisting medium, 111
Roemer, Olaus, 135
Royal Observatory, Greenwich, 13, 92-3, 140
Royal Society, 13, 65-6, 92, 94, 96, 136, 140, 143-4, 149, 151

St. Andrews, 123
Satellites, 37, 93, 113, 115
"Scholia," 107
Sewstern, 5, 71, 154
Sextant, 135, 148
Shakespeare, William, 1, 3
Skillington, 7
Smith, Barnabas, 5
Sound wave, 111
Space and time, 103
Spectrum, 53, 67, 73, 139
Stirling, James, 151-2
Stoke, 7
Stoke Rochford, 7, 45
Storey, Miss, 11
Sun, 30-1, 37, 88, 114, 116
Swift, Dean, 142

Taylor, Brook, 144
Telescope, 53-4, 60, 63, 65-67

Theology, 125-6, 146
Thin films, 74
Tides, 93-4, 111, 115-6
Transparency and opacity, 78
Turnor, Edmund, 7

Uniform circular motion, 33, 38, 48
Unity in universe, 49
"Universal Arithmetic," 140
Universities, 97
Uranus, 118

Volume, 25

Vortex Theory, 42, 112, 122, 146

Wallis, John, 25, 108, 125, 145
Wave-theory, 69, 77, 79
Whiston, William, 135, 139, 141
William III, 123, 131, 135, 137
Woolsthorpe, 2, 45, 51, 71, 82-4, 88-9, 154
Wren, Christopher, 89, 90, 96, 99, 137

Young, Thomas, 79

For Product Safety Concerns and Information please contact our EU representative GPSR@taylorandfrancis.com
Taylor & Francis Verlag GmbH, Kaufingerstraße 24, 80331 München, Germany

www.ingramcontent.com/pod-product-compliance
Lightning Source LLC
Chambersburg PA
CBHW061449300426
44114CB00014B/1908